Y0-CQC-938

Advance praise for BIOLOGY AFTER THE SOCIOBIOLOGY DEBATE

"Carmen James Schifellite has produced a remarkably suggestive study of the way(s) in which sociobiology, a newly christened field in 1975, became incorporated into mainstream college biology textbooks and curriculum. Dr. Schifellite has done a marvelous job of integrating a variety of issues surrounding sociobiology and its critics: a detailed examination of the conceptual claims made on behalf of the new science, its methodological foundation, criticisms launched by scientists and philosophers, the frequently one-sided (largely positive) way in which textbooks portrayed sociobiological ideas while deemphasizing the controversy that was involved, and the image of scientific process that was conveyed to students through exposure to this topic. A must read for all biologists, textbook authors, and educators."

—*Garland E. Allen, Professor of Biology, Washington University in St. Louis*

"An insightful account of the social construction of the increasingly popular and pervasive paradigm of sociobiology. Dr. Schifellite's critical analysis of the presentations of this paradigm in current college biology textbooks should be read not only by biology teachers, but anyone concerned about misleadingly reductive explanations of human behavior."

—*D. W. Livingstone, Canada Research Chair in Lifelong Learning and Work, Professor Emeritus, Department of Sociology and Equity Studies OISE/UT, University of Toronto*

"How does a scientific field of study develop, establish legitimacy, defend its truth claims against those who would contest them, and change over time? And how are the complexities of these changing claims and contestations best translated into scientific textbooks? Carmen James Schifellite engages these questions in his impressive account of the development of sociobiology and its successor sciences, such as behavioral ecology and evolutionary psychology. His detailed analyses of the rhetorical tools used in the textbook representations of the contested conceptual terrain of sociobiology, genetics, evolutionary theory, and the very nature of science, itself, are incisive and truly revelatory. Because the way we think about the relation between biology and human behavior has immense social consequences, it is imperative, Dr. Schifellite argues, that scientific texts foster the development of critical skills that allow readers to evaluate the truth claims of controversial scientific debates. This is a must read for anyone who cares about the quality of scientific debate and wishes to read—or write—scientific texts intelligently."

—*Susan McKinnon, Professor of Anthropology, University of Virginia*

"Carmen James Schifellite deftly takes the reader below the surface of a scientific controversy so that a higher level of intellectual discourse can take place within and about biology textbooks and beyond. A remarkably rich theoretical and practical book for biologists, sociologists, and educators."

—*John Novak, Professor of Education, Graduate and Undergraduate Program, Faculty of Education, Brock University*

Biology After the Sociobiology Debate

PETER LANG
New York • Washington, D.C./Baltimore • Bern
Frankfurt • Berlin • Brussels • Vienna • Oxford

CARMEN JAMES SCHIFELLITE

Biology After the Sociobiology Debate

What Introductory Textbooks Say About the Nature of Science and Organisms

PETER LANG
New York • Washington, D.C./Baltimore • Bern
Frankfurt • Berlin • Brussels • Vienna • Oxford

Library of Congress Cataloging-in-Publication Data

Schifellite, Carmen James.
Biology after the sociobiology debate: what introductory textbooks
say about the nature of science and organisms / Carmen James Schifellite.
p. cm.
Includes bibliographical references and index.
1. Biology—Social aspects. 2. Biology—Textbooks. 3. Biology—Philosophy.
4. Sociobiology—Philosophy. I. Title.
QH333.S35 570—dc22 2011011835
ISBN 978-1-4331-0018-5 (hardcover)
ISBN 978-1-4539-0252-3 (e-book)

Bibliographic information published by **Die Deutsche Nationalbibliothek.**
Die Deutsche Nationalbibliothek lists this publication in the "Deutsche
Nationalbibliografie"; detailed bibliographic data is available
on the Internet at http://dnb.d-nb.de/.

The paper in this book meets the guidelines for permanence and durability
of the Committee on Production Guidelines for Book Longevity
of the Council of Library Resources.

29 Broadway, 18th floor, New York, NY 10006
www.peterlang.com

Printed in Germany

For
Marie, Claire, and Dominic Schifellite,
whose love and support got me started,
but who were not able to see this book completed

Cover Illustration: This is an image of an installation created by the late Japanese designer and poster artist Shigeo Fukuda, entitled "Lunch with a Helmet on." The core of the installation is a sculpture made from welded cutlery. When a light source is shone through the sculpture, it casts the shadow of a motorcycle. The interaction of light and sculpture creates an image that is, to me, a whimsical metaphor for the interplay of genes and environment in the creation of organisms. It is one of Fukuda's many playful and clever illusions, and I am indebted to Shizuko Fukuda for permission to reproduce this work here.

CONTENTS

ACKNOWLEDGMENTS

There are many people whose efforts have been essential to the completion of this work. Some gave key feedback on an earlier version of this work: Erminia Pedretti introduced me to the nature of science debate in the science education literature; Garland Allen took time from a sabbatical to offer extremely insightful comments and critique; Jonathan Barker not only provided his valuable analysis, but also the generous donation of office space; and David Livingstone gave me crucial direction and feedback and, most significantly, stuck with me and with this work long after his institutional obligations required it of him.

I thank all of my colleagues in the Department of Sociology and beyond at Ryerson University for their support over the past three years. In particular, I want to thank Terry Gillin for his guidance; Carla Cassidy for financial and tactical support; Andre Foucault for help navigating the institutional issues around short-term disability; and Stephen Muzzatti for encouraging me to "keep writing."

I also want to thank Phyllis Korper, Caitlin Lavelle, Sarah Stack and all the people at Peter Lang Publishing who have been patient and enabling as I have dealt with health issues and the delays that arose from that situation. In addition, I thank Ann Martin and Gary Burford for their invaluable editorial contributions.

Special thanks also go to Nancy Barker, Greg Blaney, Christine Brown, Dickson Eyoh, Ronak Ghorbani-Nejad, Pablo Idahosa, Michael Kaufman, Bernard Kelly, Linzi Manicom, Susan McKinnon, John Novak, Andrew Pienkos, Steve Sabat, Robert Shenton, and Dan Yashinsky for their contributions, advice and support.

There are so many others who did not have a direct hand in this book's production but who have helped a great deal with general material and emotional contributions that made it possible for me to finish this book. Unfortunately, there are too many to list and I would not risk hurting some by naming others. Therefore, I will confine myself to simply thanking all of my Connecticut and Delhi family, GSU colleagues, Brock University colleagues, Toronto friends, Biltmore brothers, and dear friends in SchiPerPel Inc. for all of your support in turbulent times.

I cannot adequately express my appreciation to my partner, Pramila Aggarwal, who worked a triple day as breadwinner, bread maker, and mother while acting as a writing consultant throughout this process. I can only hope to be able to do the same for her. And finally, to our children, Manisha and Arun, who put up with the changes in our domestic routines and with my absences with humor and love.

INTRODUCTION

In 1975, E. O. Wilson synthesized much of the then-current work on the biology of the behavior of organisms under the banner of "sociobiology" in his book *Sociobiology: The New Synthesis* (Wilson, 1975; see also Wilson, 1980, 2000). Wilson's aim was to subsume under biology the things traditionally studied by disciplines such as ethology. The last chapter of his text was the most controversial, as it laid out the basis for applying this framework to human behavior and human society. Within this framework, those behaviors, cultures, social relations, and formations, which had been the sole purview of the social sciences, were given "genetic and evolutionary" explanations. His work generated a great deal of interest and controversy.

The most controversial elements within sociobiology have been the claims made about the relationship between genetics, evolution, and human behavior. Wilson claimed that this new paradigm could account for many complex human behaviors, and, to some extent, even the subsequent inequalities between people of different sexes, races, classes, and ethnic groups. This new paradigm also seemed able to naturalize these differences and structured inequalities and to transform them into simple biological and genetic evolutionary adaptations. Moreover, Wilson seemed as well to have had, as a general project, both the assertion of genetic and biological influences as the ultimate controllers of many human behaviors, and the relegation of the social sciences to second-class status in the articulation of human nature and the motivations of human behavior.

Sociobiology has been a major force in the organization of a paradigmatic position that advocates a strong role for the influences of genes on human behaviors. Despite the numerous critiques and controversy surrounding it, human sociobiology and its derivations have attracted scholars and money. There has been a very long and sometimes acrimonious debate around the ideas presented, especially in human sociobiology. This debate changed the course of the development of sociobiology and influenced the development of parallel disciplines such as evolutionary psychology and behavioral ecology. Through this debate, some scholars and scientists moved beyond the polarizing positions of "nature vs. nurture" and instead moved toward more complex ideas about the nature of organisms and the nature of science itself.

This book is about the sociobiology debate that began with the publication of *Sociobiology: The New Synthesis* (Wilson, 1975). It is about the content of this debate and about the effect that this debate has had on biology. It also is about the ways in which this debate has influenced some of the content in selected introductory biology textbooks. Although I discuss some of the history around the debate in Chapters 1 and 2, this is not a formal historical account. Neither is it an attempt to add new material to the work developed by critics. Rather, this material is introduced and analyzed as a way of highlighting important themes that emerged from the debate. I am interested in highlighting the important issues that emerged from the mature debate around human sociobiology, as I believe they are still important in constructing and maintaining a non-reductive discourse on the relationship between human biology and human behavior.

The articulation of these themes is intended to help future researchers develop nondeterministic formulations and also is used to inform my textual analysis of the textbooks. I also am interested in the ways in which biology textbooks have presented sociobiology and the controversy it has generated. As well, I am interested in how changes in our understanding of the nature of science theory and practice have filtered into biology textbooks and how all of these issues can be used to help foster more critical and reflective thinking within science education as we approach the difficult issues embedded in emerging biotechnologies. As is evident in Part II of this book, this opportunity has begun to be taken up in the most recent textbooks I examine.

In this regard, throughout this book I use a number of terms to refer to sociobiology and related texts. If I am speaking specifically about Wilson's work, and to work that is similar to and sympathetic to Wilson's project, I use the term "sociobiology" or "human sociobiology." If I am referring to conceptualizations that may begin or take inspiration from Wilson's organizing texts but have proliferated now into many forms in many directions, I use the term "sociobiological discourse" or "Wilsonian sociobiological discourse."[1] When I am speaking more generally about the project of creating conceptualizations that embody a strong genetic program with respect to human behavior, I also use the terms "neurogenetic determinism," "genetic determinism," and "biodeterminism."

Why the Sociobiology Debate Now?

One might ask why it is useful to study this debate, especially in light of Lewontin and Levins' (2007) observation that at this point, "sociobiology has become a term of some opprobrium in biology" (61). At the same time, developments in molecular and developmental biology have moved in unexpected directions. For example, E. Keller (2000) has indicated, with surprise, that as a critic of the Human Genome Project she had expected that

> so exclusive a focus on sequence information was both misguided and misleading....Contrary to all expectations, instead of lending support to the familiar notions of genetic determinism that have acquired so powerful a grip on the popular imagination, these successes pose critical challenges to such notions. (5)

This same sentiment is reiterated by Fausto-Sterling, who has noted that as a consequence of knowledge gained in the past decade, "developmental biologists who study the role of genes in development are busily dethroning the gene" (Fausto-Sterling, 2000b, para. 1). She also notes that

> the important story is that the search for genes that control development has shown us that our initial idea that genes control processes within an organism is wrong. Instead genes are one set of actors within a developmental system. The system itself contains all of the pre-existing contents of the cell, organ and organism....What the last decade of research on genes in development reveals is that...the system and its history control development. Genes are but one of many crucial components of the process. (para. 4)

This sentiment is upheld by Craig Venter and his colleagues, who successfully mapped the human genome and who warn that we must avoid the dual pitfalls of reductionism and determinism and the mistake of discussing human variability as we gain increasing knowledge of the human genome (Venter cited in Lerner, 2004, 4). This shift also has been accompanied by the call for more balanced conceptualizations of the relative influences of nature and nurture (Ridley, 2003).[2] Despite these developments and the fall from grace of the sociobiology name, many of the key ideas and the core of the human sociobiology project remain active. Although alternative conceptions have emerged, they are not yet dominant (Kaplan and Rogers, 2003, 5). Evidence for this comes from a number of areas.

In the late twentieth century there has been much support and interest in the academic world for the aims and ideas of human sociobiology, and support remains strong within a number of disciplines. Ruse has documented data taken from the *Science Citation Index* that indicates that between 1975 and 1995, Wilson's *Sociobiology: The New Synthesis* has been cited 2,040 times by authors (Ruse, 1999, 148). Likewise, Segerstråle's more recent search of the Wilson Index alone (as different from E. O. Wilson) yielded 13,000 entries under "sociobiology" (Segerstråle, 2000a, 314). In a survey of books in print in September 2000 using only the keyword "sociobiology," I found 150 titles published in the preceding nine years, and 80 of these had been newly published or republished in the preceding three years.[3] The focus of these titles has been very wide-ranging. In the past twenty-five years people have written about the "biology of love," the "genomics of selfishness," "altruism," "desire," and "homosexuality," to name a few.

Second, while the term "sociobiology" is in decline, the general project has been taken up by newer disciplines such as behavioral ecology and evolutionary psychology (Pinker, 2002). For example, as of August 2009, the online Human Behavioral Ecology Bibliography (HBEB) listed well over 1,000 books and articles published, in press or in process since 2000.[4] Evolutionary psychology has become an important subdiscipline and, while many take pains to try to ensure that it is not a deterministic focus, critics insist that it is falling into the same paradigmatic positions as were present in the early human sociobiology formulations (McKinnon, 2005; Kaplan and Rogers, 2003). The proliferation shows no signs of abating. These contemporary formulations cross a wide spectrum. There are new directions and more balanced accounts of the interconnections between nature and nurture; nonetheless, there are more of the same old biodeterminist formulations. For those advocating a strong determinist program, the talk may be about control by hormones or brain structures rather than by genes, but in the end these more recent formulations rely on conceptual underpinnings that are similar to Wilson's original work. One could say that the sociobiology debate is still ongoing, and, while in this book I confine myself primarily to discussions about sociobiology and related discourses that emerged around the same time, I do discuss some of the more recent developments and offshoots in Chapter 1.

A third reason to pay attention to the lessons of the sociobiology debate is that sociobiology and related determinist discourses also have had success in entering the popular imagination. A great deal of both popular and academic research and writing continues on and about sociobiology and the general issue of evolution, genetics, and human behavior. Most significantly, the resonances among determinist discourses, advancements in biotechnologies in the past twenty years, and the promises of future innovations sometimes act to reinforce and support determinist claims, especially in the popular imagination (Allen, G., 2002).

The advances in biotechnology began in the mid-twentieth century with the deciphering of the nucleic acid structures of genetic materials and with the first successful clones of amphibian organisms. This was followed in the 1970s and 1980s with early mapping of sites of significant genetic material in simple organisms and with the successful deciphering of some of the ways in which these sites operated in the production of specific proteins. In the past three decades, we have witnessed the continued technical development of genetic mapping techniques. This technical development has been coupled with research that has begun to reveal the connections between genotypes and specific diseases. Advances in the 1980s and 1990s in the deconstruction, mapping, and reconstruction of DNA led to the development of the Human Genome Project. The Human Genome Project proceeded faster than expected. It has led to discoveries of genetic markers that are implicated in diseases. Garland Allen (2002) talks about the influences of mechanistic materialism on our conception of the gene and about the influence of the Human Genome Project on reinforcing these reductionist and simplistic conceptions. Concurrent development in the biotech industry has spurred on all facets of genetic mapping and research. And, as well, pre-implantation genetics diagnosis has now begun to move toward the "engineering" of children.

We are increasingly immersed in a world in which references to genetics are being used to explain illness, health, well-being, and even behavior (Allen, 2001). Often, this is done by referencing intermediary factors such as brain morphology or hormone action, but ultimately the final arbiter of control is usually put down to genetic influences. Lippman (1991) was one of the first to warn of this process of "geneticization" of society, and she sees it as

> an ongoing process by which differences between individuals are reduced to their DNA codes, with more disorders, behaviors and physiological variations defined, at least in part, as genetic in origin. (19)

These developments in genetics, proteinomics, and the biotechnological applications of this work have contributed to the creation of a kind of "gene-talk" within many public and popular discourses today (Keller, E., 1995; 2000). In these discourses, genes and DNA are being given a preeminent status in the controlling of human biology and human behaviors (Hubbard and Wald, 1993). Lewontin (1991) talks about this phenomenon as the development of what he calls "The Doctrine of DNA." Rifkin (1998b) considers this the development of the "sociology of the gene." Nelkin and Lindee (1995) talk about the gene as having developed into a cultural icon.

> The findings of scientific genetics—about human behavior, disease, personality and intelligence—have become a popular resource precisely because they conform to and complement existing cultural beliefs about identity, family, gender and race. The promises made by scientists reflect these beliefs. Such promises express the desire for prediction, the need for social boundaries, and the hope for the control of the human future. At the same time, scientists' claims about the powers of the gene meet many social needs and expectations. Whether or not such claims are sustained in fact may be irrelevant; their public appeal and popular appropriation reflect their social, not their scientific, power. (197)

The story of Hamer's work on the "gay gene" illustrates Nelkin and Lindee's point. Kaplan and Rogers (2003) note that even though Hamer's work on the "gay gene" has "not been sustainable, the social construct of the 'gay gene' has persisted" (212).

E. Keller (1995, 2000) identifies this "gene-talk" as first, early in the twentieth century, a discourse of "gene action" and then in the latter half of the century it became the discourse of the "genetic program." This "gene talk," sociobiological discourse and the sociobiology debate itself will become increasingly important in framing the influences and effects of this ongoing geneticization process. Advances in genetic mapping technology will continue to promote the process of geneticization outlined above. And, most importantly, the development of all these technologies herald that we may be on the verge of a new eugenics movement (Allen, G., 2001). If so, we will all need to develop much more critical and evaluative capacities when negotiating these new eugenic technologies.

This new eugenics movement likely will be more consumer based than the previous movement, and it will possibly coalesce around two processes or factors. The first is an emergent trade in information about genetic traits/markers. This is the practice of selling information about traits as genetic markers through consultative processes. This is already beginning to happen (Goetz, 2007; Wallace, 2005). Trade can involve wellness counseling for the postnatal or genetic counseling for the prenatal. In this scenario, we as consumers pay for knowledge of our genetic predispositions so that we can take action to control potential disease processes and promote health and longevity.

The second process or factor is the development of an emergent trade in the sale of genetic material through a process that attempts to commodify biological traits as genetic markers. This form of commercialization has been termed "liberal eugenics" (Betta, 2006, 7). In this scenario, we as consumers would purchase traits or markers as ordered up in the creation of embryos. The idea of ordering up the color of your children's eyes, for example, may sound like science fiction right now, but this is already beginning to happen. If one looks at the history of pre-implantation genetic diagnosis, we can see that it was created to allow parents to screen their future children for serious medical conditions. In addition to this use, however, parents from around the world have already begun to use this technology for sex selection of their unborn children in countries such as the U.S., where such practices are legal (Baruch et al., 2006, 5; Kilani and Hassan, 2001). As well, donor insemination now has been transformed from a semi-clandestine and secretive process designed to help infertile couples into an open and international business with a global trade in sperm and eggs (DiLascia, 2006; Zachary, 2000). It is now possible to take what are being called "in vitro fertilization vacations."[5]

Prior to and accompanying these increases in more conscious and overt eugenic manipulations is a great deal of talk and conjecture about the influences of genes and hormones on complex human behaviors. At the same time, new technologies will increase our ability to practice eugenics—to attempt to select some of the traits of our future children (Schifellite, 2008). For example, on January 6, 2007, *The Washington Post* (Stein, 2007) reported that The Abraham Center of Life LLC, started by Jennalee Ryan, the founder of the largest adoption agency in the U.S., is producing "ready-made" embryos for purchase;

prospective buyers can review donor information such as race, education, appearance, and personality prior to purchase (Saletan, 2007). The ability to select eggs and sperm from donors has long been available to parents, but this is "the first time anyone has started turning out embryos as off-the-shelf products" (ibid). According to Robert P. George,

> this is just more evidence that we haven't been able to restrain this move towards treating human life like a commodity. This buying and selling of eggs and sperm and now embryos based on IQ points and PhDs and other traits really moves us in the direction of eugenics. (cited in Stein, 2007)

If this business flourishes, it will have profound effects on society. According to Ryan, the founder of The Abraham Center of Life LLC discussed above, there is "a demand for white babies," and "three-quarters of the DNA in [the Center's] first two batches comes from blue-eyed blondes" (Saletan, 2007). This means that consumer preference for specific physical and "behavioral" characteristics will drive the demand for donors and that consumer preference for specific physical and "behavioral" characteristics also will drive the price of the sperm and eggs that are procured. For example, Ryan is expecting that a tiered pricing system will come into effect for their embryos and that "[their] compensation is offered to those donors who have earned a post-graduate degree [or] have a unique skill, characteristic or trait" (cited in Saletan). As Saletan puts it, "PhD embryos will cost more than BA embryos." These are not just the musings of journalists. Kalfoglou et al. report that clinicians running pre-implantation genetic diagnosis (PGD) centers are beginning to fear that PGD will become seen as a tool not just for sex selection but also for other non-medical traits such as eye color, physical appearance, intelligence, and sexual orientation (2005, 492–493).

At this point, what is being offered is only a crude kind of eugenics, but soon the technology offered may claim to also offer germ-line genetic manipulation. Prospective parents will be confronted with offers to provide eugenic enhancements to zygotes or germ-cell genetic material. Some of these enhancements will likely be marketed to them as behavioral enhancements or "upgrades" for their unborn children. Prospective parents may find themselves presented with the seeming ability to create children in whom traits such as particular personality types or intelligences have been selected for. Imagine being told that you could select for attractive extroverts with high intelligence and a low likelihood of developing addictive and/or criminal behaviors. Some likely will be willing to take the chance. Some will reject these kinds of interventions outright as violating some moral, ethical, or religious principles. Still others will object to such a simplistic and deterministic linking of genetic material with complex human behaviors. This debate will happen against a backdrop in which researchers attempt to track homosexuality genes and "empathy" chromosomes and look for cognitive frameworks that support the genetic and evolutionary roots of complex human behaviors such as altruism and female mate-selection.

To be sure, many ethicists and social scientists have raised the alarm and argue against allowing such eugenic manipulation (Solberg, 2005; Rifkin, 1998b; Maranto, 1996). It may not be possible, however, to stop the development of a "positive" eugenics that is used in attempts to modify the intelligence, personality, physical appearance, or sexual orientation of our own offspring (Dahl, 2003; Baker, 2000; Gardner, W., 1995). We have lived through the past fifty years in which there has been an intense debate and many vocal proponents and opponents of the idea that it is even possible to manipulate complex human behavior and performance by manipulations at the genetic level. Those promoting the possibility of eugenic technologies will find an important ally in a popular imagination that has been influenced by the context of geneticization discussed above. Also, continued successes in discovering the connections between genotype and disease will help legitimize more reductionist formulations about the relationship between genes, cells, systems, organisms, and environment. These successes have emboldened and will continue to embolden some to make dubious, but popular, proclamations about the role of genes in producing personality, criminality, sexual orientation, altruism, selfishness, aggression, etc.

Ordinary citizens and politicians already are being asked to debate questions on a variety of subjects not confronted before by human societies. In the future, we likely will be called upon to make decisions about selecting the traits of our future children, to make decisions about what kinds of medical treatment we will receive, and to make judgments about the use of social resources. We will be asked to make these judgments based upon what biological and genetic "theories" tell us about the potential "natures" of groups of people. Questions will arise around the appropriateness of genetic screening for predispositions to what have been considered complex human behaviors such as mental illness, personality traits, and intelligence (Paul, 1994).

How will we make these decisions? What should we allow? Are these claims even possible? In all these examples and in a host of others yet to emerge, citizens and politicians will be asked, based upon the advice of "experts," to create policies that are informed by controversial claims about our natures, diseases, and potentials. Implicit in the marketing of this kind of technology will be the assumption that there are genetically controlled biological processes that determine, in a relatively straightforward way, complex human behaviors. Developments in biotechnology, eugenics technologies, and the apparent success of the Human Genome Project have increased legitimacy for a paradigmatic shift in the debate over what influences human behavior.

The Focus of This Book

It is important that these technological changes happen within a discourse that debates not just the ethical issues created by these choices but also within a discourse that continues to render problematic the whole relationship between genetics and human behavior. This discourse needs to continue to challenge the idea that we will be able to select for not just physical characteristics, but also for characteristics or complex behaviors such as intelligence, personality, and sexual orientation. This is a discourse that emerged out of the sociobiology debate, and it is important to make available and to continue to develop the arguments and issues that have arisen as the debate has proceeded and positions have been clarified.

These arguments and issues signal conceptual areas that must be addressed both within science education and by the lay public as a new eugenics movement develops. It is crucial to begin to foster education and debate among teachers, students, scientists, and the lay public about this issue. It is important that biology teachers and scientists have the necessary background knowledge about the issues so that they can effectively convey the issues to students and the lay public. We all must become conversant enough in this debate to be able to make informed judgments about the theories and claims that will continue to surround us and about the eugenic "products" that may be marketed to us. These judgments require that we develop evaluative skills at many levels. As such, this work utilizes theory developed within the sociology of science and also relies on theory developed in hegemony studies, discourse theory, and textual analysis.

In coming to grips with the sociobiology controversy, teachers and textbook authors also must engage the questions that arise around the nature of science knowledge and method. As I discuss in Chapter 3, numerous authors have begun to do work on this topic within science education, and continued focus within science education that questions our assumptions about the nature of science can only improve our abilities to critically evaluate positions that arise from scientific controversy. All of this will not be easy, however. Critics of Wilsonian sociobiological discourse must be able to engage the debate at the levels of genetic theory, evolutionary theory, and theory about the nature of science as well as at the level of sociobiological formulations.[6] It is my hope that this book will make a contribution to the ways we think about this issue and to the ways we understand the underlying theories that legitimize and underpin the linking of genetics and human behavior. It is my aim that this work will help to improve and clarify the debate around the relationship between biology and behavior for teachers, students, and the lay public in general.

This book is divided into two parts. In Part I, I describe and analyze the central issues that have emerged from the sociobiology debate. In Chapter 1, I examine the development of human sociobiology, its basic tenets, and its related disciplines. In addition, I chronicle the major issues that critics raised in the first fifteen years of the debate, and I examine recent developments in human sociobiology and related fields and the work of recent critics. In this chapter, I also examine the recent emergence of alternative and non-reductive formulations that take into account the many biological, psychological, and social determinants of human behavior. These are complex debates that are ongoing and unsettled.

In Chapter 2, I cover what I think are the most important points of difference that separate those advocating a strong biological determinism from their critics. I also examine the ways in which both sides understand key concepts such as reductionism and biological traits and the ways in which these understandings also influence how both sides conceptualize organisms as their objects of study and the nature of the life and social sciences themselves. In effect, sociobiology has been part of a larger debate that has proceeded within the life sciences and within science studies. In Chapter 2, I also examine the ways in which Wilson and others constructed legitimacy for sociobiology and the areas of contention that emerged as the debate matured. In conjunction with this analysis of the construction of the legitimacy of this discourse, I also examine the obstacles that critics face in attempting to undermine the legitimacy of simple and reductive presentations about how biology, in general, and genes, in particular, influence human behavior, and I examine the ways in which the critique against these kinds of formulations can be sharpened. A number of questions follow from this line of investigation. For example, "Why has a strong genetic model maintained prominence in the popular imagination?" And

"How can a nuanced and mediated position that takes both the biological and the social into account in any talk of the relationship between biology and human performance and behavior be strengthened?" In Chapter 3, I examine the way in which the nature of science itself is conceptualized. In this chapter, I cover some of the debates that have developed in science studies and within science-education literature. Also in Chapter 3, I look at the way in which one's thinking about the nature of science and epistemology in general relates to one's position on sociobiology.

In Part II of this book, I look at representations of sociobiology in university-level introductory biology textbooks (see Chapter 4 for the complete list). In Chapter 4, I introduce my method of analysis, the textbooks that I chose, and the reasons for these choices. I could not examine every introductory biology textbook in print in the depth that I needed to for this study. Instead, I chose six texts for this analysis. Four were published in the mid-1990s and two are later editions of two of the original texts I selected. These latter two texts were published in 2005 and 2008. In this way, I have been able to present a picture of these textbook representations and the observable changes in these representations that occurred over two decades.

In Chapters 5 through 8, I examine these introductory biology textbooks in detail to see if and how they cover sociobiology and its offshoots, and how they cover the sociobiology controversy and some of the themes that developed in the mature debate. In this regard, in this textual analysis, I do more than examine how the textbooks portray sociobiology. I also examine their presentations of themes that emerged in the sociobiology debate that are controversial issues in genetic theory, evolutionary theory, and the nature of science itself. This analysis has helped me to understand the ways in which sociobiology, genetics, and evolutionary theory are structured in biology textbooks and the importance of presenting issues around the nature of science in these textbooks. This analysis has also given me a way both to gauge the influence of human and nonhuman sociobiological discourse and to gauge the ways in which critical thinking and controversy are handled by the texts. Finally, in Chapter 9, I summarize my findings on these textbook representations. I also discuss the implications of my findings in both parts of this book, for science educators and for all those who oppose strong biologically determinist formulations of human behavior and support the development of more balanced and more biologically and epistemologically modest formulations. I am not a historian or a philosopher and I do not claim in this book to have written definitively in either of these areas. I am a sociologist and educator, and as such this work is sociology of knowledge focused on the sociobiology debate and also a form of sociology of knowledge as textual analysis involving particular introductory biology textbooks. As such, it contains both the strengths and weaknesses of such an undertaking.

Last, I would like to add that, while I am critical of some of the ways specific material is presented in some of these introductory biology texts, I have the utmost respect for the huge task that all these authors have taken upon themselves. These introductory texts are massive books of more than one thousand pages each, and they cover the very vast knowledge base that we call "biology." All of these authors put in great effort to support the best aims and aspirations embodied in scientific practice and to make biology exciting and fascinating for students. They all succeed at this, and I thoroughly enjoyed the considerable time that I spent examining them in detail.

Notes

1. This notion of "Wilsonian discourse" is similar to Keller's (1995) notion of discourse that she borrows from Hacking and to Laclau and Mouffe's (1985) notion of a hegemonic discourse. In all cases, the basic concept is of a set of ideas and practices that limit and define the boundaries of reality and the boundaries of conceptualization for the given subject. I discuss this in more detail in Chapter 2. And, although I use the name Wilsonian discourse, others and especially Dawkins (1976, 1979, 1982, 1986, 1995) also have contributed to this discourse.

   Nonetheless, this is not a monolithic conception. In Chapter 1, I also discuss the various branches and directions, which, to me, have been both positive and negative, that have developed in the past three decades. In the latter half of the 1980s and into the 1990s the critiques moved away from more personal and overtly political attacks and began to focus on some of the core issues that were dividing the camps. There are the Wilsonians who envisioned the possibility of a human sociobiology with what I would call a strong genetic program. There are also those who see the possibility of integrating investigations in the life and social sciences in ways that allowed for the uniqueness of each level of analysis and the uniqueness of organisms (especially human) without the need for an over-determining geneticism.
2. For more discussion on this point, see the subsection on "alternative positions" in Chapter 2.
3. This list has included *The Darwinian Heritage and Sociobiology* (Smillie, van der Dennen, and Wilson, 1999); *Sociobiology and Bioeconomics* (Koslowski, 1999); *Sociobiology and the Arts* (Bedaux and Cooke, 1999); *Marx and Sociobiology* (Huaco, 1999); *Marxism and Human Sociobiology* (Chang, 1994); *The Biology of Love* (Janov, 2000); *Human/Nature: Biology, Culture and Environmental History* (J.P. Herron and Kirk, 1999); *Living with Our Genes* (Hamer and Copeland, 1998; *The Moral Animal* (Wright, 1994); *Crisis in Sociology:*

The Need for Darwin (Lopreato and Crippen, 1999); *The Science of Desire: The Search for the Gay Gene and the Biology of Behavior* (Hamer and Copeland, 1994); *Why Sex Matters: A Darwinian Look at Human Behavior* (Low, 2000); *Evolutionary Psychology: The New Science of the Mind* (Buss, 1998); *The Darwin Wars* (A. Brown, 1999); *Defenders of the Truth: The Battle for Science in the Sociobiology Debate and Beyond* (Segerstråle, 2000a).

4. This bibliography is maintained by Kermyt G. Anderson at the University of Oklahoma and can be found at http://faculty-staff.ou.edu/A/Kermyt.G.Anderson-1/HBE/index.html#2008.
5. A quick Internet search will reveal a number of businesses selling this service. Two sites I found were ivfvacation.com and MedicalTourismCo.com. It also is being discussed in various forums and blogs on the Internet.
6. A PhD dissertation by Macdonald (2000) provides disturbing evidence that about two-thirds of pre-service science teachers he tested had a poor understanding of basic evolutionary concepts.

PART I

The Sociobiology Debate

· 1 ·

THE SOCIOBIOLOGY DEBATE

An Overview

Historical Antecedents

The discipline called sociobiology burst onto the scene in 1975 with the paradigm-organizing publication of E. O. Wilson's text entitled *Sociobiology: The New Synthesis* (Wilson, 1975). The attempt to use concepts taken from evolutionary theory and genetics to understand what motivates behavior in humans and nonhumans did not begin with this book, however, nor only recently. It has its roots in a number of fields. Wilson's nonhuman sociobiology can be seen as part of a relatively uninterrupted project of studying animal behavior informed by evolutionary and genetic theory that has ranged from at least Darwin through the ethologists. His human sociobiology resumed a debate, however, about the influences of genes, natural selection, and the environment on human behavior that had been halted, or at least interrupted, in the post–World War II years.

The roots of the various attempts at constructing a biology of the human and nonhuman social space can be traced to Darwin, Spencer, and Malthus (Jones, 1980). Ever since Darwin first published his work on evolutionary theory, he and others have speculated about the ways in which evolution and genetics affect not just physical characteristics but behavior as well. The early twentieth century saw the initial development, in biology, of what is called the modern synthesis. At this time, Mendel's work was rediscovered; various researchers uncovered the location of the genetic material, and the chromosome and gene were named. In this development, evolutionary theory was merged with the newly developing science of Mendelian genetics and, later, population genetics. This newly developing science of genetics was seen as a strong ally and support for the evolutionary paradigm.

Three things came together in biology in the early twentieth century. First, the modern synthesis with its evolutionary and genetic foci came to dominate biological discourse and research (Sapp, 1990a, 107–109). Along with this development came not only a renewed interest in theorizing the genetics of human and nonhuman behavior but an interest in human eugenics as well. This eugenics movement wanted not only to understand where behavior came from but, more importantly, to selectively breed more "ideal" behaviors into the human population and to breed out more "undesirable" behavior (G. Allen, 1994; Kaplan, 1994; Kevles and Hood, 1992). Given these developments, it is not surprising, for example, to see A. G. Keller, in 1915, writing *Societal Evolution*, in which according to Turner, Maryanski & Giesen (1997), he

> sought to use Darwinian concepts as replacements for Spencer's ideas about evolution....Instead Keller proposed that notions of variation, selection, transmission and adaptation should be used in the analysis of social evolution. (23)

In conjunction with this developing theory, one also sees the development of eugenics societies, staffed by reputable scientists who advocate controlling human breeding through counseling and forced sterilization (G. Allen, 1994). E. Keller (1992b) cites historian Diane Paul when discussing the consensus among virtually all geneticists, who, by the 1920s, came to believe in a strong genetic determinism "concerning the role of heredity in the determination of intellectual, psychological and moral traits" (283). This determinism was the theoretical underpinning for the eugenics movement of the early twentieth century (Paul, 1984). E. Keller (1992b) adds to Paul's argument the idea that this genetic determinism included the "belief in the power of genes to mold the character of human beings" (284).

This theorizing developed into what became known as the modern nature/nurture debate, in which factions divided around the relative weight accorded to either genetics or environmental influences in the conceptualization of the determinants of behavior. To be sure, the debate around the innate and environmental causes of human behavior is very old. Ancient Greek philosophers debated these points, as did Darwin and Lamarck in the nineteenth century. In terms of the twentieth-century debates, Keller contends that after World War II, in the revulsion against Nazi eugenics and racist eugenics in the U.S. and U.K.,

> the direct link between genetics and its eugenic implications that had earlier been so visible, and so powerfully motivating, was no longer politically tolerable. (E. Keller, 1992b, 285)

Against this backdrop following World War II, nature came to be seen as the deciding factor influencing human nature and human behavior. And as Keller shows,

> . . . [A]ll reference to the human uses to which a knowledge of genetics might be put was abandoned...[and] human behavior could be claimed as a free zone while at the same time the science of genetics—and with it, confidence in the genetic determination of everything but human behavior—could prosper....Simply put, it was nurture, not nature, that was seen as conducive to the kind of unfettered development imagined possible by a victorious and "free" republic. And in the absence of a strong public stance to the contrary by geneticists, the general optimism of the time inclined both popular and academic assumptions about the relative importance of nature and nurture (at least in the realm of human behavior) to undergo a decisive shift. In the mood that came to prevail, anything seemed possible, given the right environment and the right kind of nurturing. (E. Keller, 1992b, 285–286)

Weingart and Segerstråle see the UNESCO statement on race drafted in 1949 and revised in 1951 as a turning point after which "the climate in society and academia had 'officially' shifted in favor of environmental explanations" (Wiengart & Segerstråle, 1997, 75–79). I think that E. Keller is correct in seeing that in the conceptual vacuum that followed, the strong nurture or environmental position became dominant.

Within genetics, an attempt at a balanced nature/nurture position was exemplified in the work of Dobzhansky. In addition to his feeling that genetics could not encapsulate behavior, he counseled against reductionist explanations in favor of those who tried to grasp the complexities involved in the mechanisms of genetics and natural selection (Dobzhansky, 1956, 1964, 1967). His position can be seen in these passages taken from his well-known article, "Of Flies and Men":

> Human nature is, then, not unitary but multiform; the number of human natures is almost as great as the number of humans!

> The demonstration of the genetic uniqueness of individuals only opens, rather than solves, the problem as far as behavioral and social sciences are concerned. . . .
>
> [Likewise], there are no genes "for" behavior as there are no genes for the shape of one's nose; the problem is more subtle. (Dobzhansky, 1970, 429–430)

Yet, even as the nurture side predominated in the years following World War II, genetics was growing in stature within biology such that "by the late 1960s, genetics had moved to center place in the life sciences" (E. Keller, 1992b, 286). And it was a particular kind of genetics that became hegemonic.

A revival of genetic determinism and the re-linking of genetics and behavior had slowly been building momentum when *Sociobiology: The New Synthesis* was published. E. Keller (1992b) cites a comprehensive survey conducted by the National Academy of Sciences in 1968 entitled *Biology and the Future of Man*, which dealt with future expectations of the biological disciplines (287). In the last chapter of the nine-hundred-page document, "the question of 'man' emerges with force, and with it the exemption of human behavior that had heretofore been so carefully maintained was dissolved" (287). While this is in many ways a heralding text in the re-unified gene-based behavior paradigm, it was not alone. We can also see publications from the 1950s and 1960s that begin to develop disciplinary foci to answer the questions as to how and to what extent evolution and genetics control human and nonhuman behavior.

At this time, we can see a revival of interest taken up by the eugenics movement of the early twentieth century. For example, in 1963, *Brains of Rats and Men: A Survey of the Origin and Biological Significance of the Cerebral Cortex*, originally published in 1926, was republished (Herrick, 1963). Also, a symposium held by the Eugenics Society in 1967 produced a compilation edited by Parkes and Thoday (1968) entitled *Genetic and Environmental Influences on Behavior*. In the foreword to this edition, the authors note how the success of their recent symposium had encouraged the council of the Eugenics Society, through the medium of the newly established Galton Foundation, to launch *The Journal of BioSocial Science* in 1969. Also, in 1967, a compilation entitled *Genetic Diversity and Human Behavior* was published (Spuhler, 1967). It arose out of the Burg Wertensteen Symposium No. 27 and was entitled "Behavior Consequences of Genetic Drifts in Men." The symposium focused on human behavioral genetics.

In general, in the 1950s and 1960s, there was a resurgence of interest in the connecting of genetics and behavior that seemed to gather force in the mid-1960s, when a number of discipline-building publications began to appear. Hall wrote an article entitled "The Genetics of Behavior" for *The Handbook of*

Experimental Psychology in 1951 (Hall, 1951). Caspari wrote an article entitled "The Genetic Basis of Behavior" in 1958 (Caspari, 1958). Fuller and Thompson wrote the first *Behavioral Genetics* text in 1960, and an edited compilation entitled *Behavioral Genetics: Method and Research* was published in 1969 (Manosevitz, Lindzey and Thiessen, 1969). The first paper in this last publication was by Lindzey (1969), entitled "Genetics and the Social Sciences." In it, he noted prophetically: "a number of recent developments suggest a growing interest in the implications of genetics for the social sciences" (3).

Within this new wave of publications around genetics and behavior, there are those who adopt what could be called a strong program of genetic determinism and those who adopt a weaker, more nuanced or restrained model reminiscent of the aims in Dobzhansky's work. An example of the latter can be seen in a colloquium sponsored by the Werner Reimer-Stiftung in 1977 (two years after Wilson's *Sociobiology: The New Synthesis* was published), which produced a collected volume of papers entitled *Human Ethology: Claims and Limits of a New Discipline* (von Cranach, Foppa, Lepenies and Ploog, 1979). In this volume, the editors raise a number of questions about the extent to which findings from animal ethology can relate to human culture and society. It is clear that participants seem to want to provide a bona fide space for both the social and the biological in attempting to understand complex human behaviors. For example, in the introduction, they ask, "Are there typically human forms of behavior that cannot be dealt with adequately in terms of (animal) ethology?" (xiii)

What we have developing in the 1960s and 1970s is an attempt to re-balance the nature/nurture debate so as to include a place "at the table" for "nature." Ultimately, those advocating a strong biological determinism become prominent. It was into this intellectual breach that Wilson jumped, with his own agenda of rebalancing the emphases in the nature/nurture debate in favor of a stronger biological and hereditarian determination on human and nonhuman behaviors.

Wilson was not the only one working on these issues, but his formulation became the most popular. There were a number of books "trying to achieve a similar kind of synthesis of the new developments in the field of social behavior" (Segerstråle, 2000a, 85). Segerstråle cites five books all published not long before *Sociobiology: The New Synthesis*, all of which dealt with similar ideas and none of which received the attention that Wilson's book received.[1] For example, A. Brown notes that M. T. Ghiselin's (1974) book, *The Economy of Nature and the Evolution of Sex*, "fizzled out almost unremarked" (A. Brown, 1999, 49). And Steen and Voorzanger (1984) found that while Robert Hinde's (1974)

Biological Bases of Human Social Behavior covered the same issues as Wilson did in *Sociobiology: The New Synthesis*, his book was "completely disregarded in many sociobiological texts." Yet it was Wilson's synthesis that ignited the imagination and incurred the ire of both the popular and scientific audiences.

The Immediate Roots and Basic Concepts of Sociobiology

Numerous people had been working on a general theory of gene selectionism and the relationship between evolution and behavior. Wilson takes the concepts of "kin selection" and "inclusive fitness" from W. D. Hamilton (Kitcher, 1987, p.77; Wilson, 1975, 56; see also Hamilton, 1964, 1996), and the concept of "reciprocal altruism" from R. L. Trivers (A. Brown, 1999, 50; Kitcher, 1987; Wilson, 1975, 58; see also Trivers, 1971, 1985). He gets the general idea of selfish genes and the use of games theory from R. A. Fisher and J. B. S. Haldane by way of Hamilton as well (see Kitcher, 1987, 79; Hamilton, 1964, 1996). Hamilton (1964), influenced by Price (Hamilton, 1996), had developed the ideas of inclusive fitness in the mid-1960s and applied them to the question of altruistic behavior. Trivers (1971) had written in the early 1970s on the sex differences in "parental investment" in birds. Maynard Smith (1964) had written on the concept of kin selection. Segerstråle (2000a) summarizes this historical coalescence well:

> . . . [W]e have Fisher, Haldane, and Sewall Wright, two British and one American, the background architects of gene-selectionist thinking. Bill Hamilton can make Britain proud, but his close collaborator, George Price, was an American, who had come to live in England. Later, Maynard Smith collaborated with Price on the ESS [Evolutionary Stabile Strategies]. Hamilton, again, partly traces his thinking to an early paper of George Williams (and D. C. Williams), in yet another transatlantic influence. Williams (1966), in turn, was crucial for Dawkins (and seemingly Wilson, too), but Williams got activated by reading Wynn-Edwards. Trivers' early papers were an important inspiration for Dawkins (Dawkins, 1989, 298); Dawkins, together with Wilson, became vehicles for spreading Trivers' and Hamilton's ideas (Dawkins also promoted the notion of an ESS). And these ideas, again, had a complex relationship with ethology—partly originating from ideas in this field, partly influencing them, and partly being influenced by them. (89–90)

Wilson's enterprise was both synthetic and creative. His major contribution was to take concepts developed by others and weave them into a new discipline that connected genetics and evolution with the study of behavior in human and nonhuman societies (A. Brown, 1999; Kitcher, 1985; Segerstråle, 2000a). The concepts of inclusive fitness, kin selection, and reciprocal altruism opened new avenues for thinking about the ways selection can operate on behavior. For Wilson, "selection can be said to operate at the group level....If selection operates on an individual in any way that affects the frequency of genes shared by common descent in relatives, the process is referred to as *kin selection* [author's emphasis]" (Wilson, 1980, 50). That is, selection can, and does, operate on individuals in ways that can affect groups and, in particular, one common "grouping" is composed of genetic relatives; this grouping process is called kin selection and is connected to the theory of inclusive fitness.

Inclusive fitness essentially means that instead of simply viewing the fitness potential of an individual and his or her genes, one must consider the effects a behavior has on the individual's genetic relatives. The notion of inclusive fitness that Hamilton developed includes "the idea that our genes are spread into subsequent generations by our relatives" (Kitcher, 1985, 81). According to Wilson (1975), inclusive fitness is

> [t]he sum of an individual's own fitness plus all its influence on fitness in its relatives other than direct descendants; hence the total affect of kin selection with reference to an individual....[And, kin selection is] the selection of genes due to one or more individuals favoring or disfavoring the survival and reproduction of relatives (other than offspring) who possess the same genes by common descent. (586–587)

Wilson traces the general idea of kin selection back to Darwin's "idea of natural selection operating at the level of the family rather than of the single organism" (Wilson, 1980, 56). Ultimately, kin selection and inclusive fitness are combined with theory and mathematics developed in evolution biology, population genetics, and games theory. This combination is used to explain why specific behaviors such as altruism have been maintained in nonhuman societies, even in cases in which the specific behavior appears to reduce the fitness—that is, the survivability of the individual.

In other words, if an organism does something that endangers its life for the sake of its siblings, cousins, nieces, or nephews, there is a point at which the personal sacrifice is equivalent to the "good" it does in increasing the survivability of its shared genes. Thus, there would be a tendency to preserve a genome that favored the production of altruistic acts toward close relatives because the preservation of one's close relatives preserves significant portions of the same genome. To avoid a Lamarckian tinge, genetic "coding" to recognize kin and genetic "coding" to act altruistically must be postulated so that selection pressures have something to act on.

For Wilson, this new synthesis solves a particularly vexing problem in evolutionary theory—the problem of altruism. Entomologists had long observed apparently altruistic behavior in insects. Yet it was difficult to imagine how behavior that reduces the survival of an individual could continue to be selected for. Would not those individuals and ultimately their genes that determine this behavior disappear? In order to understand how a seemingly maladaptive behavior can flourish, the concepts of "inclusive fitness" and "kin selection" are crucial. The concepts of kin recognition and natural selection allow the sociobiologist to view altruism as an act under genetic control or influence that is "intended" to promote the survival of the genome of a close genetic relative—who the altruist is able to "recognize" as such. Or, to be more precise, altruism is a trait that has been selected for and that is "genetically encouraged," because in the long run, the trait will increase the fitness of the genes that "cause" or increase the likelihood of the altruistic act. Human individuals who practice altruism toward kin and reciprocal altruism toward non-kin will thus be selected for.

In addition to kin selection and inclusive fitness, Wilson also takes from Trivers the idea of reciprocal altruism. With this concept, Trivers uses mathematics developed in games theory and evolutionary theory to show that genes that increase the likelihood of altruistic behavior toward even non-relatives (reciprocal altruism) can have the effect of increasing the long-term survivability or personal fitness of the altruist. Likewise, in this case, in order to avoid Lamarckian explanations, the genetic disposition to act altruistically must be present. Selection pressures must have something to act on. If indeed there is genetic control (or influence) affecting the display of such altruistic behavior, then it would likely be selected for.

For the kinds of nonhuman, insect, and bird societies that provide important data out of which these theories develop, other conceptual components are required. In the case of insect societies or bird flock behaviors, it is necessary to assume that insects and birds are not consciously choosing to be moral or to preserve the genes of their relatives. Instead, they are acting out of biological impulses and ultimately genetic influences that can be subject to selection processes. With the theories of kin selection, inclusive fitness and reciprocal altruism, it is also necessary, in the nonhuman world, to postulate a biologization of kin recognition and a biologization of altruistic behavior. Kin selection needs to involve the ability to recognize one's relatives for it to work. The organism needs to be able to distinguish between relatives and non-relatives.

Ultimately, this recognition ability must have an underlying genetic basis. In order for selection to operate, there must be something to select. Immediately, it is a behavior that is selected for, but ultimately, for Wilson, it is a genotype that is being selected. Wilson's sociobiological argument, especially for creatures such as insects and birds but fostered as well for other animals, requires that both kin recognition abilities and altruistic or self-sacrificing behavior be somehow genetic in origin and thus selectable. Sociobiological explanation assumes that these behaviors are biological characteristics or traits with fairly direct linkages to genes that behave in fairly straightforward Mendelian fashion.

Wilson then extends his analysis to cover a wide range of social behaviors. He lays out his theoretical foundations in the introductory chapters of *Sociobiology: The New Synthesis*. He then goes on to a section entitled "Social Mechanisms," which has chapters with topics such as "The Development and Modification of Social Behavior," "Communication: Origins of Evolution," "Aggression," "Dominance Systems," "Roles and Castes," "Sex and Society," "Parental Care," and "Social Symbiosis" (Wilson, 1975, 1980, 2000). All of these topics, for the most part, deal with nonhuman behaviors and societies. The behaviors he discusses, such as aggression, dominance, caste formation, sex selection, parental care, "rape," and "homosexuality" in nonhumans are also constructed as biological traits or characteristics that behave in a similar Mendelian fashion.

One set of behaviors examined by Wilson are the differences in male and female reproductive strategies in humans and nonhumans. According to Wilson, sociobiological theory predicts that given the very different "investments" both sexes are biologically committed to make in the rearing of offspring, one will find that through natural selection, very different behaviors have evolved for the sexes to optimize the reproduction of their DNA. For males, promiscuity and the impregnation of as many females as possible are the behaviors that will most likely increase the survival chances of the males' genome. For females, serial monogamy with a male committed to investing resources into the rearing of offspring is the behavioral strategy selected for, which will most ensure the propagation of the females' genome.[2] Rather than seeing these behaviors in human males and females as animated by conceptual, emotional, or existential influences, within the sociobiological framework these behaviors are seen as traits with a strong evolutionary, genetic, and biological basis.

It is fair to say that in Wilson's synthesis, concepts such as inclusive fitness and behaviors such as kin selection, altruism, and reciprocal altruism must be grounded in a strong hereditarian framework and in an evolutionary model that favors genes as the important "units of selection" and level of analysis. This latter position resurrected an earlier debate on units of selection. Are populations, the behaviors of individuals (phenotypes), or genotypes the proper arena in which to focus? In this "units of selection" debate Wilson and Dawkins' work placed sociobiology clearly in the "gene" camp. Consequently, organisms and their behaviors are not the real focus of sociobiology, but rather genes, as Wilson indicates:

> In a Darwinist sense, the organism does not live for itself. Its primary function is not even to reproduce other organisms; it reproduces genes, and it serves as their temporary carrier.... But *an individual organism is only a vehicle* [author's emphasis]...the organism is only DNA's way of making more DNA. (Wilson, 1975, 3)

In other words, the complex behavior of organisms has been selected for through evolution such that individual organisms act in ways that increase the chances that their genomes will propagate and be successful. In this circle of causality, individuals appear to act so as to increase the fitness of their genomes and these genes are "selfish" (Dawkins, 1977). They act to ensure their own propagation through control of the organism. Wilson's discussions covered a very wide range of nonhuman behaviors and societies from insects to higher primates. Wilson's most controversial pronouncements, however, were those made about human social behaviors.

Human Sociobiology

Wilson did not stop at nonhuman societies, and in the first and last chapters of *Sociobiology: The New Synthesis*, which make up only five percent of the text, he discussed human behavior and human societies. These are the pages that have drawn much of the attention and criticism of readers. For Wilson (1975), sociobiology "is also concerned with the social behavior of early man and the adaptive features of organization in the more primitive contemporary human societies" (4). Sociology has a "largely structuralist and non-genetic approach. It attempts to explain human behaviors primarily by empirical description of the outermost phenotype and by unaided intuition. . . ." (4).

In contrast, "Sociobiology is defined as the systematic study of the biological basis of all social behavior" in animal and human societies (Wilson, 1975, 4). And, while Wilson admits that human behavior is under many complex forces, he sees a strong hereditarian connection among all creatures and holds that "perhaps it is enough to establish that a single strong thread does indeed run from the conduct of termite colonies and turkey brotherhoods to the social behavior of man" (1980, 63).

In the first and last chapters of *Sociobiology: The New Synthesis*, Wilson lays out areas of human social behaviors that would seem fertile ground for sociobiological study. The human qualities that he focuses on are those that, to him, "appear to be general traits of the species" (1980, 272). He talks about barter and reciprocal altruism, gendered sexual behavior, bonding, the division of labor, role playing, polytheism, communication, culture, ritual, religion, ethics, esthetics, territoriality, and tribalism (1980, 272ff). He also holds that "genes promoting flexibility in social behavior are strongly selected at the individual level" (1980, 273).

Wilson (1980) also suggested a mode of analysis that entails comparing hunter-gatherer societies with other primate species, so that "it might be possible to identify basic primate traits that lie beneath the surface and help to determine the configuration of man's higher social behavior" (275). Using this method, he concludes that it is very probable that early man shared a number of traits with nonhuman primates. These included a group size of one hundred or less, male dominance over females, territoriality, game playing, prolonged maternal care, socialization of young, and extended maternal/offspring relationship—especially between mothers and daughters (293).

He does not confine himself to only these behaviors, however. In his writings on human sociobiology, he also ventures into the realms of moral philosophy and epistemology. Wilson says in *Sociobiology*: "[S]elf knowledge is constrained and shaped by the emotional and control centers in the hypothalamus and limbic systems of the brain" (1975, 3). And, since these systems have "evolved by natural selection" the correct field of study in human sociobiology should include biological explanations of things such as "ethics and ethical philosophies, if not epistemology and epistemologists, at all depths" (3). Wilson has a lot to say about human altruism and moral philosophy. The human sociobiological argument removes behaviors such as altruism from the domain of moral philosophy. Instead it locates them as potential traits that can be read through a basic Mendelianism and a strong genetic program.

Sociobiology generated tremendous controversy when it first appeared, especially its attempts to transfer work done on insects and animals to primates and humans. Had Wilson kept his focus on the nonhuman world and avoided any talk of a human sociobiology, he likely would have drawn much less criticism and much less popular interest. Segerstråle examined several factors that may have contributed to the particular success of Wilson's book. She cites Dawkins' observation that Wilson "devotes more space to the comparative study of social systems and their correlation with environmental variables" which he saw as missing from the other books (cited in Segerstråle, 2000a, 85). She also cites Maynard Smith's observation that without the last speculative chapter on humans he doubted "the book would have achieved the fame—or the sales—that it did." (cited in Segerstråle (2000a), 86)

The Effects of Sociobiology

Wilson's synthesis is important for a number of reasons. It establishes both the disciplines of human and nonhuman sociobiology as well as initiating the popular debate around human sociobiology. This is not the only way this synthesis could have occurred, and this version certainly goes in directions others working in the field of behavioral studies might not have gone. It is a synthesis that clearly resonated, however, with a great many inside the scientific community and with the lay audience. It synthesized a great deal of research that had been going on over the previous twenty years in animal behavior (Wilson, 1975; see also Wilson, 1980, 2000). Wilson's aim was to subsume under biology the things traditionally studied by disciplines such as ethology. Etkin (1981) notes that sociobiology emerged as a fusion of population genetics and field ethology (52). It not only synthesized this research, however, but did so in the context of creating a new, or, as some would argue, renewed paradigm. This new synthesis was strongly hereditarian in its understanding of the "roots" of behavior in nonhumans and strongly but—critics have argued—confusingly hereditarian in its presentation of the determinants of human behaviors. His work generated a great deal of interest and controversy. The last chapter of his text was the most controversial, as it laid out the basis for applying this framework to human behavior and human society. Wilson claimed that this new paradigm could account for many complex human behaviors and the subsequent inequalities between people of different sexes, races, classes, and ethnic groups. This new paradigm also seemed able to naturalize these differences and structured inequalities and to transform them into simple biological and genetic evolutionary adaptations.

Also out of this synthesis came the claim that a new science or discipline was being presented that would subsume the terrain and theory of social science under a new sociobiological approach. The aim for sociobiology, especially human sociobiology, was hegemonic in that it claimed to provide a better way of understanding human societies than the social sciences currently provided. Within this framework, those behaviors, cultures, social relations, and formations, which had been the sole purview of the social sciences, were given "genetic and evolutionary" explanations.

Critics were very caustic in denying that this "withering away of social science" would occur. But even one of Wilson's strongest critics, Philip Kitcher (1985) has held that sociobiology is not, as some critics claim, a unidimensional approach to behavior. He believes that to bolster this defense,

> sociobiologists can point to a variety of theoretical ideas—the notion of inclusive fitness, the application of game theory to the understanding of fitness relations, the use of optimality models to analyze fitness. (77)

The particular genius and attraction in Wilson's *Sociobiology* are in its synergies—both theoretical and ideological.

Sociobiology Since *Sociobiology: The New Synthesis*

What has happened to the project of developing a Wilsonian-type human sociobiology since these early formulations? The answer seems to be that, for some, it has undergone many transformations, and for others it has stayed the same. Wilson and other like-minded theorists did not stop with the publication of *Sociobiology: The New Synthesis*. Rather, it opened the door for a huge number of publications from those critical of its project and from those wishing to expand the ideas further. This development has not all happened exactly within the confines of Wilson's original formulations. Wilson went on to write *On Human Nature* (1982), an elaboration of his work in human sociobiology. Moved by critics, he also went on to co-author, with Charles Lumsden, *Promethean Fire* (Lumsden and Wilson, 1983) and *Genes, Mind and Culture* (Lumsden and Wilson, 1981), which are their attempts at both a mathematicization of the principles of human sociobiology and also an integration of culture through the idea of gene-culture coevolution. As Degler puts it in *In Search of Human Nature*:

> There was a time when sociobiology was seen by friends as well as critics as imperialistic, seeking to subsume all of human experience under the Darwinian Imperative. [But] Even Ed Wilson by the early 1980s was talking interaction. (1991, 310)

Critics were not impressed by this later work, however.[3] Also, Richard Dawkins published *The Selfish Gene* (1976), *The Extended Phenotype: The Gene as the Unit of Selection* (1982), and *The Blind Watchmaker* (1986). In all of these, he sought to refine ideas about the evolution of human and nonhuman behaviors presented in *Sociobiology: The New Synthesis*. His more recent work (2006) has continued in this vein. Hrdy, among others, continued developing more feminist inspired ideas about male and female strategies of reproduction in a number of books (Hrdy, 1977; 1981; Hausfater and Hrdy, 1984). Others, such as W. D. Hamilton (1996), Trivers (1985), R. D. Alexander (1979; 1987; 1988; 1990), and Maynard Smith (1988, 1989) continued working and publishing within academe rather than in the more popular press. At the same time, older related disciplines, such as behavioral genetics, have continued to develop; and new and related disciplines such as evolutionary psychology have emerged. Some take seriously the idea of considering socio-cultural and biological factors equally, while others stick to the earlier Wilsonian formulations.

Recent Developments Within the Sociobiological Family

There have been alternative biosocial formulations that have arisen both within the larger sociobiology movement and outside it.[4] According to Blute (2003), currently this research takes three directions:

- human behavioral ecology/evolutionary psychology/sociobiology emphasizing biological (gene-based) evolution,
- sociocultural evolutionism emphasizing sociocultural (social learning or meme-based) evolution, and
- gene-culture coevolution emphasizing their interaction (gene-culture coevolution).

The emergence of evolutionary psychology is descended from the original sociobiology discourse but at the same time Badley (1931) used the term in his much earlier book. According to van der Dennen, Smillie, and Wilson, "such terms as 'behavioral ecology' and 'comparative psychology' have come to be preferred to cover Wilson's broad field of cross-species comparisons" (1999, xii). As I indicated above, a connection between behavioral genetics and psychology goes back to the early nineteenth century. It is not surprising, then, that we should see the development of "evolutionary psychology" in the conceptual wake of sociobiology.[5] A seminal text in this endeavor is *The Adapted Mind*, a collection of papers edited by Barkow, Cosmides, and Tooby (1992). In this work, the authors wish to introduce "the newly crystallized field of evolutionary psychology to the wider scientific audience" (3). The authors point out in the introduction of this text that the volume is "centered on the complex evolved psychological mechanisms that generate human behavior and culture" (3). They also point out that this new field "unites modern evolutionary biology with the cognitive revolution in a way that has the potential to draw together all the disparate branches of psychology into a single organized system of knowledge (3).

Barkow et al. (1992) hope that this can be managed while respecting the integrity of the many different disciplines and levels of analysis.

> The second goal of this volume is to clarify how the new field...supplies the necessary connection between evolutionary biology and the complex, irreducible social and cultural phenomena studied by anthropologists, sociologists, economists and historians.
>
> To understand the relationship between biology and culture one must first understand the architecture of our evolved psychology. Past attempts to leap-frog the psychological—to apply evolutionary biology directly to human social life—have for this reason not always been successful. Evolutionary psychology constitutes the missing causal link needed to reconcile these often warring perspectives. (3)

Mulder, Maryanski, and Turner (1997) hold hope for the synthesis that can arise out of the merger of anthropology and evolutionary psychology. They cite the weakness in sociobiology that it did not specify well the way in which "evolutionary processes might influence human behavior" (36). For them, evolutionary psychology might allow sociobiologists to "extend their analytical objective from that of the actor's acts to that of his or her goals, beliefs, and motives" (36). Critics contend, however, that instead of this, much of evolutionary psychology has tended to reproduce the strong emphasis on genetics (Lewontin and Levins, 2007; McKinnon, 2005; Kaplan and Rogers, 2003).

The second model that Blute highlights—sociocultural evolution—focuses on the ways that culture evolves in its own right. The idea here is that the development of culture can be studied in much the same way that we can study the evolution of organisms. This idea was first suggested by Dawkins (1976) in his concept of the "meme" as a unit of cultural evolution, and it can be seen in the work of Boyd and Richerson (1985). Dawkins developed the concept of the "meme," to refer to a cultural unit analogous to genes that we internalize in our thoughts and language and that we "pass on" and preserve through socialization in a "meme pool." [6] This work emphasizes "that because culture is transmitted, it can be studied employing the same Darwinian methods used to study genetic evolution" (Turner et al., 1997, 30). It is studied, however, as independent of any genetic or other biological influences. This idea of sociocultural evolution has also been incorporated into coevolutionary models that emphasize both biological and cultural influences working independently and in concert in human history.

According to Blute, coevolutionary models are the third group that has emerged in the wake of *Sociobiology: The New Synthesis*, and these are a serious attempt to integrate genetics, evolutionary theory, and biochemistry with cultural and social influences in ways that allow autonomy and integrity for all elements in this conceptual effort. This group wishes to develop a more complex and balanced theorization of the interaction of genetics and culture. These more recent coevolutionary approaches

> only assert that isomorphisms between the processes of biological and cultural transmission can be the basis for constructing theoretical models. Such an approach does not aver that cultural transmission is reducible to genic processes. (Turner, Maryanski, and Giessen, 1997, 30)

This is an attempt to develop serious models that include factors such as culture, cognition, choice, and social influences when considering the determinates of human social behavior.

A noteworthy example of this kind of work is the collaboration edited by Peter Weingart, Sandra D. Mitchell, Peter J. Richerson, and Sabine Maasen (1997), entitled *Human by Nature: Between Biology and the Social Sciences*. In it, Peter Weingart explains that this book first saw light as

> [a] project to bring together biologists and social scientists for an academic year at the Center for Interdisciplinary Research (ZiF) to discuss their differences and agreements, and perhaps their mutual agnosticism, the working title of the project was meant to be provocative: *Biological Foundations of Human Culture*. (Weingart, 1997, vii)

Human by Nature is the product of that year of work and, as Weingart points out,

> the problem then, was to find a common methodological ground to bridge the gap between different disciplinary approaches to the study of specific aspects of human culture.... We avoided biological or sociological reductionism.... We adopted the model of "integrative pluralism" as its methodological strategy. (viii)

Human by Nature is a wide-ranging compendium that gives weight to the idea that twenty-first-century discussions of biology and society can be less reductionist, more complex, and more likely to build in an expansive space for the social. It contains very useful and informative overviews of disciplinary efforts at theorizing about the interactions of biology, society, culture, and cognition.

In this collaboration, Turner, Maryanski, and Giesen (1997) describe what they term an alternative to sociobiology (echoing Gould's observation twenty years ago), which emphasizes "that cultural phenomena operate in terms of Darwinian principles" (29). In so doing, they want to make a distinction between earlier versions of this kind of reasoning put forward by R. D. Alexander (1977), Dawkins (1976), and Lumsden and Wilson (1981, 1983) and the more recent coevolutionary approaches that have emerged out of anthropology and evolutionary psychology. Early on, Baldwin and Baldwin (1981) in *Beyond Sociobiology* called for the development of a balanced biosocial theory that could incorporate natural selection and natural learning. The biosocial approach also has been developed by Walsh (1995) and Buss (1994, 1999; Blute, 2010), among others. Within the biosocial field, there also are "more genetic" and "more social/cultural" formulations.[7]

Turner, Maryanski, and Giesen describe both interactionist and dual inheritance models. W. H. Durham (1991) has developed an example of the interactionist model in *Co-evolution: Genes, Culture and Human Diversity*. As he describes it, his aim in proposing a coevolutionary perspective is, among other things,

> to suggest that, despite the complexity of their relationships, natural selection (or, more precisely, genetic selection) and cultural selection do tend to co-operate in the evolution of the attributes that are adaptively advantageous for some, it not all, of their bearers. I call this theory "co-evolution" to emphasize that, for this group of beneficiaries...genetic selection and cultural selection have generally harmonious, parallel influences in guiding the evolution of human diversity. (Durham, 1991, 419)

Turner, Maryanski, and Giesen also highlight the work of Cavalli-Sforza and Feldman (1981), as another example of what they feel is a successful interactionist model. The dual inheritance models offer a more independent conceptualization of the effects of biology and culture.

Mulder et al. summarize the current state of coevolutionary work as follows:

> For some (e.g., Durham, 1991), biological evolution is but one type of a more general process that also includes cultural evolution. For others (e.g., Boyd and Richerson,

> 1985, 1990), biological processes can provide the conceptual leads to develop distinctive models for understanding the evolution of traits in sociocultural systems. There also are differences in how much the biological and cultural inheritance systems influence each other. For some who remain sympathetic with sociobiology (e.g., R. D. Alexander, 1979; Lumsden and E. O. Wilson, 1981, 1983), much sociocultural inheritance is circumscribed by biology. For others much less committed to sociobiology (e.g., Boyd and Richerson, 1985; Cavalli-Sforza and Feldman, 1981), the two systems of inheritance—the genetic systems of biology and the traits of sociocultural systems—are distinctive, although understandable with similar models emphasizing variation and selection processes. (1997, 37–38)

This coevolutionary research is the most likely work that will be able to provide a bridge between the social and biological disciplines and produce new forms of knowledge that do not attempt to simply override the traditional areas of social research, while at the same time adding to our understandings of the complex interactions between the biological and the social realms. In addition to all of the transformations and developments that grew out of all of this work, the development of sociobiology—especially human sociobiology—drew a firestorm of critiques from biology and from the social sciences. The substance of these critiques is what I turn to next.

Critical Reactions to Sociobiology

Critical reaction to Wilson's *Sociobiology: The New Synthesis* came quickly. Numerous critics of his human sociobiology arose. Some challenged it as bad science; others deemed it pure ideology. In addition, the sociobiology debate focused a good deal of dialogue and criticism on issues and controversies within evolutionary theory (Segerstråle, 2000a). The first set of criticisms of sociobiology came from a group of academics, scientists, doctors, and graduate students who came to be known as the Sociobiology Study Group of Sciences for the People. Their first piece, called "Against 'Sociobiology,'" appeared in *The New York Review of Books* on November 13, 1975 (E. Allen et al., 1978). A second piece, entitled "Sociobiology—Another Biological Determinism," published under the name Sociobiology Study Group of Sciences for the People, appeared in 1976 (see Sociobiology Study Group of Sciences for the People, 1978).

A number of overviews and edited compilations on the subject, both pro and con, were published soon afterward, the most notable being *The Sociobiology Debate* (Caplan, 1978); *Sociobiology Examined* (Montague, 1980); and *Sociobiology: Beyond Nature/Nurture?* (Barlow and Silverberg, 1980). Critiques of sociobiology and biodeterminism in general continued during the 1980s.

Perhaps the most complete and devastating critique of most aspects of human sociobiology has been Philip Kitcher's *Vaulting Ambition* (Kitcher, 1985). Steven J. Gould and Richard Lewontin, who were key figures in the Sociobiology Study Group and colleagues of Wilson's at Harvard, went on to write more extensive critiques. There are, in addition, other important critics, such as Garland Allen, Steven Rose, Hilary Rose, Richard Levins, and Ruth Hubbard.[8] These early critiques, which date from 1975 to the mid-1980s, are summarized in the following sections.

Human vs. Nonhuman Sociobiology

Most of the early critiques focused on the problems that critics saw with Wilson's attempts at human sociobiology. The Sociobiology Study Group's critiques were based on the premise that Wilson simply did not and could not make a case for the extrapolation from insect and animal sociobiology to humans. Indeed, most of the criticism of Wilson and others who write in this vein is centered on their blurring of the lines between conceptualization in nonhuman sociobiology and the attempts at extrapolation to human behavior. These early critics also focused on the apparent political impact and political agenda of Wilson's discussions of human sociobiology. Only later did more substantive critiques and discussions emerge that focused on some of the controversial ideas that sociobiology borrowed from genetic and evolutionary theory.

Sahlins (1977) and, later, Kitcher (1985) provide systematic and in-depth critiques of the human sociobiology project, and both make a distinction between Wilson's human and nonhuman sociobiology. Both authors begin from the premise that there seem to be two sociobiologies in Wilson's work. One is a scientifically plausible and nuanced conceptual framework that considers only nonhuman societies. The other is a widely speculative adventure into discussions of the genetic causes of human behavior. Sahlins and Kitcher both contend that Wilson's nonhuman sociobiology has merit and is worthy of serious consideration, and both argue that his human sociobiology is seriously flawed, simplistic, crude and ideological.

Sahlins called the attempts at human sociobiology "vulgar sociobiology" (Sahlins, 1977) and Kitcher called it "pop sociobiology" (Kitcher, 1985). Kitcher defines "pop sociobiology" as the enterprise

> which consists in appealing to recent ideas about the evolution of animal behavior in order to advance grand claims about human nature and human social institutions.... (Kitcher, 1985, 15)

Sahlins (1977) defines what he calls "vulgar sociobiology" as "the explication of human social behavior as the expression of the needs and drives of the human organism, such propensities having been constructed in human nature by biological evolution" (3). Few critics challenged the need for a determinist formulation of behavior when Wilson discussed insect, avian, and even some mammalian social behaviors. E. Allen et al. (Sociobiology Study Group of Sciences for the People, 1978), however, took exception to what they perceived as Wilson's importing this model to human behavior. They held that "while evolutionary analysis provides a model for interpreting animal behavior, it [did] not establish any logical connection between behavior patterns in nonhuman and human societies" (1978, 262).

While it might be conceivable that there are similar behaviors and etiologies between primate and human behaviors, Allen et al. are critical of what they saw as Wilson's unsubstantiated movements from what might be conceivable "to what is" (262). In all cases, the authors are critical of the attempt to apply conceptual models developed for study of animal behavior to human behavior, culture, and society. Both authors also hold out the possibility that some of the best work on the behavior of insects and other nonhuman animals may be important and useful. For Kitcher, many of the charges against sociobiology in general "are overblown." Nonetheless, he contends that "the points made by the critics are relevant both to pop sociobiology and to *some* work in the sociobiology of nonhuman animals" (1985, 183).

Within this critique of "pop sociobiology" or "vulgar sociobiology," there are a number of consistent themes. Critics charge that there were a number of methodological and presentation problems, including anthropomorphism, "ad hoc" arguments, imprudent speculation, unwarranted assumptions, extrapolations from animal to human societies, a lack of evidence, and the conscious omission of competing ideas from Wilson's work. Critics also charged that *Sociobiology: The New Synthesis* was reductionist and determinist in scope. In addition, some critics saw the book as a political work and charged Wilson with intentionally trying to normalize contemporary social inequalities around race, class, and gender and to influence social policy. As time wore on, critiques also began to emerge that took exception to some of the interpretations Wilson put on certain issues in genetics and evolutionary theory. Critics and defenders alike began to realize that fundamental issues were being debated, and critics especially began to realize that what they were calling "reductionism" was in fact a complex constellation of interpretations. Untangling this "charge" is the focus of the remainder of this chapter.

"Reductionism": A Complex Problem for Critics

One of the most consistent and cogent critiques of determinist human sociobiology is that it is reductionist. But what exactly does this mean? A number of years ago, I attended a series of workshops on the history of genetics in the twentieth century. At one of the sessions, a prominent social historian of science off-handedly referred to a very important figure in the development of molecular genetics and in evolutionary theory as a "reductionist." Another historian of science present at this session, who had spent a great deal of time studying the work of this scientist, intervened in a forceful and heated manner in defense of the maligned scientist, claiming that the charges were preposterous. I thought I knew what the speaker meant by the charge of reductionism in relation to the larger discussion of theories that linked human behavior and human genetics. I also thought I understood that the defender was considering a lifelong body of work that he felt could not be dismissed with a one-word description.

One of the things I took away from this exchange is that the charge may sometimes be given too cavalierly and in place of what should be a rather detailed analysis of a series of missteps in an argument. I also learned that very few academics and scientists want to be seen as "reductionists" in the sense that the critic had meant but are trained and happy to be reductionists in a very different sense. Reductionism, within the sociobiology debate, has become what Raymond Williams (1976) called a "keyword." According to Williams,

> keywords are significant binding words....Certain uses bound together certain ways of seeing culture and society....Certain other uses seemed to me to open up issues and problems...of which we all needed to be very much more conscious. (cited in Keller and Lloyd, 1992, 4)

Keller and Lloyd (1992) have compiled a list and analysis of "Keywords in Evolutionary Biology" in their book by the same name. The authors see these keywords "as indicators of the ongoing traffic *between* social and scientific meaning and, accordingly, between social and scientific change" (4–5). Both R. Williams and Keller and Lloyd see keywords as having more than one meaning and they see these meanings shift as a result of the interplay between changes within social and natural science disciplines and changes in the larger social framework. They also argue that it is crucial that we try to understand the shifting meanings of keywords in debates. Also, Keller and Lloyd argue that it is the consistent attempt within the natural sciences to do the kind of paring down of the meanings of keywords that distinguishes the scientific enterprise and "reaps distinct cognitive benefits" (Keller and Lloyd, 1992, 3).

In the rest of this chapter and in Chapter 2, I will discuss the ways in which the charge of reductionism as directed by critics of sociobiology is really a charge about a complex set of processes employed by proponents of sociobiology. These processes turn human activity into abstract concepts, and they rely on specific interpretations of genetics and evolutionary theory to foster their credibility and legitimacy. According to critics, the Wilsonian discourse on human sociobiology employs a neopositivist and reductive method, a basic Mendelianism, what Rose calls a kind of ultra-adaptationist Darwinism, what E. Keller calls "the discourse of gene action," an overly elastic concept of "traits" or "characters," and a process of reification. Critics see these elements as woven together to form a genetic determinism that underlies and fortifies the sociobiology discourse.

Simple Mendelianism and the Discourse of Genetic Action

Human sociobiology placed itself squarely within the protective arms of two of biology's most canonical discourses—genetics and evolutionary theory. It borrows selectively from both of these disciplines. Critics charged that Wilson borrowed a simple Mendelianism that he put to use in the development of sociobiology, and that this position may have limited value, especially when speaking of complex things such as human behaviors. As Hubbard (1982) points out, "Most traits do not follow Mendel's laws" (71) and

> genes have been invoked as the originators of specific traits as well as of major structures and functions of organisms. They are called upon to explain the orderly transformations that occur during development and aging. And they are said to be decisive for long-term changes during evolution and speciation. Much of this rests on assertion and has no underlying observational basis. (72)

For critics, Wilson employs this Mendelianism as if it can explain the phenomena he studies. The bind for critics is that attacks on the genetic roots of sociobiological reductionism also can appear to be attacks on Mendel, the occupier of a canonical place in genetics and in all of biology. This is an issue that I will discuss in more detail in this chapter, as part of an examination of the ways in which sociobiological discourses have attempted to build legitimacy.

At the same time, critics see that the Wilsonian sociobiological discourse borrows heavily from what E. Keller calls "the discourse of gene action" and "the discourse of genetic programs." This discourse is not unique to sociobiology, however. According to E. Keller (1995), what developed as genetics in the early twentieth century was first a "discourse of gene action" that later was transformed into a discourse of the "genetic program." Keller articulates this discourse in the following way:

> A belief long-standing among geneticists (and one that has acquired greater currency in recent years for the public at large) is that genes are primary agents of life: they are the fundamental units of biological analysis; they cause the development of biological traits; and the ultimate goal of biological science is the understanding of how they act. (3)

She believes that this discourse took hold more strongly in the United States than in Europe and that its adoption or rejection has influenced the ways biologists understand the processes of development and embryology. She sees this discourse of gene action as "the hallmark of the American school of Morganian genetics" (11). E. Keller (2000) contends that in the mid-twentieth century this discourse was transformed, but not fundamentally altered, to become a discourse of the "genetic program." She contends that this discourse was first introduced by Jacob and Monod.

> It introduced a new metaphor for thinking about development, one with distinct advantages over the earlier notion of gene action. The metaphor of a program allowed for gene interactions in ways that the earlier metaphor did not; it resonated powerfully with recent developments in computer science; and, most important of all it could encompass the new work on gene regulation. But Jacob and Monod's innovation was not simply the proposal of a program for development; it was their proposal of a program entirely contained within the genome. It was, in other words, the notion of a "genetic program" (80).

While Keller feels that we are now moving toward a more complex and less determinist understanding of the relationships between genotypes, cells, organisms, and behaviors, these discourses still influence, in determinist ways, conceptions in genetics and in the popular imagination. Sociobiology discourse clearly adopted the discourse of gene action and of the genetic program as part of its core principles and legitimacy. This overdetermined sense of gene agency can be clearly seen in sociobiological formulations of the role of genes in controlling behavior as it relates to adaptation and evolution.

Adaptationism and Sociobiological Discourse

Critics charged that Wilson borrowed from evolutionary theory what Rose calls an "ultra-darwinism" focused almost exclusively on the operation of adaptation in evolution. In their first published criticism of Wilsonian human sociobiology in 1975, the Sociobiology Study Group had observed that Wilson relied on a strong adaptationist program that bolstered his lack of evidence for things such as "conformer genes" and "genes favoring spite." "Despite there being no evidence for these genes" for Wilson, what exists is adaptive, what is adaptive is good, therefore what exists is good" (E. Allen et al., 1978, 261). Again in a second paper E. Allen et al. (Sociobiology Study Group of Sciences for the People, 1978) contend that

> the assertion that all human behavior is or has been adaptive is an outdated expression of Darwinian evolutionary theory, characteristic of Darwin's 19th-century defenders who felt it necessary to prove everything adaptive. It is a deeply conservative politics, not an understanding of modern evolutionary theory, that leads one to see the wonderful operation of adaptation in every feature of human social organization. (286)

Segerstråle (2000a), who is a supporter of Wilson, corroborates the idea that Wilson's particular formulation of sociobiology needs a strong adaptationist position.

> [Wilson] clearly found the assumption helpful that much of animal behavior could be described by optimization models. *Optimization theory, in fact, provided Wilson's longed for heuristic for an integrated sociobiology*. In order to be able to develop an integrated sociobiology at all, Wilson needed the heuristic assumption that social behaviors represented evolutionary optima. (106)

Lewontin and Gould both launched critiques at this ultra-adaptationist version of evolutionary theory. Segerstråle (2000a) claims that Lewontin and Gould's political attacks on Wilson were only a smokescreen for the real battleground that they saw as this extreme adaptationism. In her view, the political attacks were a "Trojan horse" designed to cover what was really at issue. According to Segerstråle (2000a) and Ruse (1999), both Gould and Lewontin admit to being trained by strong adaptationists and to adopting strong adaptationist positions early in their careers. But as time went on they began to feel disenchanted with this position. Both Segerstråle and Ruse characterize Lewontin as a scientist concerned with what Ruse calls "epistemic values"—or the scientific values of "truth seeking" (Ruse, 1999, 32). Segerstråle describes

"Lewontin's general scientific attitude" in the following way:

> 1) arguments should be correct rather than simply plausible; 2) correctness is more likely to be obtained by the experimental method than any other process; 3) speculation about past evolution can be at the most plausible, never proven; therefore, it is not scientifically fruitful; 4) big generalizations are almost sure to be incorrect because of the complexities involved in evolutionary processes; 5) therefore, the most scientifically sound thing to do is to concentrate on prediction, use experimental method, and ask restricted questions. (Segerstråle, 2000a, 105)

This method thus seems antithetical to the adaptationism program of which Lewontin is critical and that Lewontin sees in Wilson's work. Lewontin calls this approach the "adaptationist program" that "assumes without further discussion that all aspects of the morphology, physiology and behavior of organisms are adaptive optimal solutions to problems. . . ." (cited in Segerstråle, 2000a, 102).

Gould as well finds fault with the strong adaptationism that he sees in Wilson's and Dawkins' work. According to Ruse (1999), Gould's movement away from orthodox Darwinian adaptationism came as he prepared to write *Ever Since Darwin* (136). This movement led him to develop theories involving punctuated equilibria, in which rapid change caused by macro-mutation and nonadaptive events could be seen to be at least as significant as evolutionary processes as uniform progressive adaptation by natural selection. For Gould (1984),

> evolution is a balance between internal constraint and external pushing to determine whether or not, and how and when, any particular channel of development will be entered. Natural selection is one prominent mode of pushing, but most engendered consequences of any impulse may be complex, non adaptive sequelae of rules in growth that define a channel....Natural selection does not always determine the evolution of morphology; often it only pushes organisms down a preset, permitted path. (cited in Ruse, 1999, 141)

And further

> I believe that the methodological flaws in human sociobiology are serious enough to incapacitate its central claim of strong genetic constraint imposed by natural selection for specific, adaptive behaviors. (Gould, 1980b, 284–285)

It seems clear that Lewontin, Gould, and others engaged in this debate for two reasons. In general, they do not agree with the notion that all change is progressive and adaptive. Some is accidental, not especially functional, and even serendipitous. Also, they engaged in this debate within the context of critiquing sociobiology because they see this "Panglossianism" in which everything is "adaptive" and "as it should be" as a key foundation in the kind of sociobiology developed by Wilson.

The combination of a simple Mendelianism coupled with the discourse of genetic action and a strong adaptationist program all go to creating within sociobiology a conception of evolution that can appear to be very teleological and in which organisms appear as lumbering robots. Critics contend that such misleading simplifications have led sociobiology into a process of reification that amounts to a self-sustaining teleology in which the influence of the social is purged.[9]

Reification: Reductionism as a Self-Sustaining Teleology

Early on in the debates Stephen Rose (1980), in an article in *Race and Class* entitled "'It's Only Human Nature': The Sociobiologist's Fairyland," used the term "reification" to analyze the process used by Wilson to "disappear" the social from his arguments. Rose's term for the process—reification—is useful because it conceptualizes reductionism as a process that occurs in the context of both internal and external ideas and discourses. It is a better conceptual tool for getting at the complex process embodied in a sophisticated determinist argument.

In the mid-1980s Lewontin, Rose, and Kamin (1984) did begin a more developed analysis of the structure of reductive sociobiological argument. In *Not in Our Genes* the authors sketch four "errors of description" in the sociobiological argument. The first is arbitrary agglomeration, in which a decision is made as to the phenotypic level of analysis under consideration. The second is "the confusion of metaphysical categories with concrete objects—the error of reification" (248). The third error is that metaphors are taken for reality and forgotten as metaphors, and the fourth is the "conflation of different phenomena under the same rubric" (250). In 1995, Rose published another piece in *Nature*, in which he developed a more thorough and systematic analysis of the structure of Wilson's and Dawkins' analyses. And in 1997 he published the book *Lifelines*, in which he devotes more effort to a systematic analysis of the structure of what he calls "ultra-Darwinist" arguments. One consistent thread, which runs through the critiques, is best summarized by Rose:

> Neurogenetic determinism, I argue, is based on a faulty reductive sequence whose steps include reification, arbitrary agglomeration, improper quantification, belief in statistical "normality," spurious localization, misplaced causality, dichotomous partitioning between genetic and environmental causes, and the confounding of metaphor with homology....[N]o individual step in this process is inevitably in error, it is just that each is slippery and the danger of tumbling very great. (S. Rose, 1997, 279)

Critics such as Rose see this process of reification as a crucial step in reductionist arguments; the process is defined in the following quote:

> Reification converts a dynamic process into a static phenomenon. Violence is the term used to describe certain sequences of interactions between persons, or even between a person and their nonhuman environment. *That is, it is a process* [author's emphasis]. Reification transforms that process into a fixed thing—*aggression*—which can be abstracted from the dynamically interactive system in which it appears and studied in isolation, as it were, in the test tube. This is the thinking that has led to regarding aggression as a phenotypic character, to be analyzed by the modern counterparts of Mendelian methods. (S. Rose, 1997, 280)

Rose has outlined a number of things going on in this process. One of the most important events is the transformation of a process, or set of social interactions and relations, into an abstracted concept. This transformation begins with the process of identifying universally observed behaviors. Rose has pointed out that human sociobiologists tend, in their arguments, to start from observed behaviors and proceeds to biologize them.

> First, what the sociobiologists do is to take aspects of human behavior and attempt to abstract certain common features from them. A mother protecting her baby from attack, a doctor risking his or her life in a cholera epidemic, a soldier leading a doomed assault on a machine gun post, all express "altruism."...Note how this trick works, by taking a process with a dynamic and a history of its own involving individuals and their relations, that is a social interaction, and isolating out an abstract, underlying fixed thing, or "quality." (Rose 1980, 162)

He goes on to describe the outcome of this process:

> What happens in this approach is that the fixed and reified property becomes attached to an individual rather than emerging from a situation. It is the individuals who then become aggressive, altruistic, intelligent and so forth, and the *same property* becomes manifest in different circumstances.
>
> Having abstracted—reified—aspects of a social interaction into uniform quali-

ties of an individual participant in that interaction, the next fallacy of sociobiological thinking is to quantify the quality. (S. Rose, 1980, 164)

In the third part comes "the appeal to biological 'evidence,'" and with the fourth "one says 'just suppose' that any particular human or social character was biologically—genetically—determined, then what would the consequences be?" (165–166)

In this critique, the determinist is seen as starting from what Rose calls a universally prevalent human behavior and ends up by discussing this behavior as a trait or character that has a biological basis and ultimately a genetic basis or cause. Once he does that, he is then able to fit that character or trait into established Mendelian models of inheritance and to extrapolate on its likely adaptive benefits. A key point in this process is the point at which a behavior, cultural practice, or social institution becomes a biological character or trait.

The British Society for Social Responsibility in Science Sociobiology Group went further and summed this process up well in a 1984 essay entitled "Human Sociobiology" (BSSRS, 1984a). I paraphrase their summary of Wilson's argument as the following:

(a) If a behavior is universal, then it is likely to be under genetic control.
(b) If a behavior is said to have a biological advantage, then it is likely to be genetic in origin.
(c) If a behavior has social advantage, then it has biological advantage. If so, it is then selected for and appears as a product of adaptation.
(d) Human society can be explained in terms of the aggregation of individual characteristics. (BSSRS, 1984a)

This selection process is essentially teleological. As Burian (1978) has noted,

in general, we have very little idea whether a specifically described behavior is under genetic control or not. If we start from arbitrarily chosen behaviors, *even if they are of selective and ecological significance*, we have no real basis for supposing that those behaviors are traits. (381)

Yet, once that leap is made, critics contend, the rest of the biologization follows. Wilsonians then simply borrow a simplistic Mendelianism from genetics and an ultra-adaptationist Darwinism from evolutionary theory.

The difficulty for critics is that all of these individual elements and interpretations have standing within genetics and evolutionary theory. The challenge for critics is to elucidate how these specific concepts are borrowed from genetics and evolutionary theory and woven into a "credible" determinist discourse. Indeed, these interpretations are contested and controversial. And so, as one expands the critique of reductionism, the critic is taken to debates outside sociobiology and to fundamental, controversial and unsettled issues within biology itself. This selective borrowing from the most dominant disciplines within biology lends legitimacy to the sociobiological project and strengthens the credibility of the process of reification outlined above while it "off-loads" the conflict to a larger conceptual setting. In part, because of these elements, a thorough undermining of the legitimacy of sociobiological discourse and related determinisms has eluded critics. It is to this issue of legitimacy in scientific discourse in general and sociobiological discourse in particular that I now turn.

Notes

1. The anthropologist Sherwood Washburn [1978] mentioned three such books: John Alcock's *Animal Behavior: An Evolutionary Approach*, Jerram Brown's *The Evolution of Behavior*, and Eibl-Eibesfeldt's *Ethology: Biology of Behavior* in its second edition, all published in 1975. George Barlow's (1991) candidates for competing syntheses were Brown's book and Michael Ghiselin's *The Economy of Nature and the Evolution of Sex* (1974). Others believed that Robert Hinde's *Biological Bases of Human Social Behaviour*, published just a year before Wilson's, received much less attention than it deserved (Segerstråle, 2000a, 85).
2. This work on sexual selection and sex-related mating strategies evoked a feminist critique from Hrdy (1981) in *The Woman That Never Evolved*. Hrdy's work looks at females' use of multiple partners, orgasm, and concealed ovulation as part of strategic female reproductive strategies. According to Maasen (1997), her formulation is more "female centered within adaptationist reasoning: Important ingredients of her evolutionary account are 'sexual selection,' 'anisogamy,' as well as 'competition among assertive, dominance-oriented females,' as central principles of primate life" (97).
3. Kitcher (1985) writes of the mathematical attempts at gene-culture interaction in *Genes, Mind and Culture* (Lumsden and Wilson, 1981): "*Genes, Mind and Culture* is an extreme example of a certain type of work. Complex mathematics is used to cover up very simple—often simplistic—ideas. What is irritating, and occasionally amusing, about these uses of mathematics is that they serve to disguise the poverty of the thought" (Kitcher, 1985, 393).

   Ruse notes that the mathematical work done with Lumsden was widely dismissed and works such as *On Human Nature* (Wilson, 1982) and *The Selfish Gene* (Dawkins, 1976) came to be seen as popularizations rather than serious contributions to evolutionary science (Ruse, 1999).
4. Although I would say this is a small movement within social science, it spans a wide range of perspective, from heavily environmental positions such as *The Biosocial Nature of Man* (Montagu, 1956) to more biologically based formulations such as *Biosociology* (Walsh,

1995).

5. J. H. Badley used the term in his book entitled *The Will To Live: An Outline of Evolutionary Psychology*, which was published by Allen in 1931.
6. For Dawkins, at some point in human history memes can become more influential and significant than genes (Dawkins, 1976). Many kinds of coevolutionary and isomorphic approaches have been developed in the last 25 years.
7. Walsh (1995) credits Mazur with coining the term "biosociology" in 1981 (2).
8. Gould continued his general critique of biological determinism, race science, and sociobiology in a series of books and articles (Gould, 1976, 1977, 1980a, 1980b). Lewontin went on to co-author *Not in Our Genes* with Steven Rose, a noted neurobiologist and critic of sociobiology, and psychologist Leon Kamin (Lewontin, Rose, and Kamin, 1984). Levins and Lewontin also went on to write *The Dialectical Biologist* (Levins and Lewontin, 1985). Steven and Hilary Rose, as well, have continued to edit critical compilations of deterministic and reductionistic accounts of human nature and human behavior (S. Rose, 1982a, 1982b, 1997; H. Rose, 1994a; H. Rose and S. Rose, 2000).

   In the 1990s, however, there were fewer critiques or examinations of sociobiology specifically but more general critiques of genetic determinism. The most notable among them were *Biology as Ideology: The Doctrine of DNA* (Lewontin, 1991); *Exploding the Gene Myth* (Hubbard and Wald, 1993); *Challenging Racism and Sexism: Alternatives to Genetic Explanations* (Tobach, Rosoff, and Fooden, 1994); *Inside and Outside: Gene, Environment and Organism* (Lewontin, 1994); and *Reconstructing Biology: Genetics and Ecology in the New World Order* (Vandermeer, 1996). Interest and critical publications seem to be rising at the start of the new millennium, however. A new crop of critical and reflective works have emerged recently, including *The Darwin Wars* (A. Brown, 1999); *Defenders of the Truth* (Segerstråle, 2000a); *The Triple Helix* (Lewontin, 2000a); *It Ain't Necessarily So: The Dream of the Human Genome and Other Illusions* (Lewontin, 2000b); and *The Century of the Gene* (E. Keller, 2000).
9. Even accepting the potential usefulness of some of the core sociobiological concepts in describing some animal behaviors, what appears to get left out of discussions is any reference to existentially based behaviors on the part of some animals and all humans. For research in this area see, for example, Clutton-Brock et al., 1999; Chase and DeWitt, 1988; and Chase, Weissberg, and DeWitt, 1988.

· 2 ·

CONSTRUCTING LEGITIMACY AND CREDIBILITY

The impact of critics on research in human sociobiology has been contradictory. In the introduction to this work, I provided evidence to show that sociobiological discourse in both its strong and weak genetic programs is still flourishing. A. Brown (1999), in *The Darwin Wars*, corroborates the fact that Wilson's work can be seen as both a successful and not-so-successful effort. For example, Brown finds that in the elucidation of the core principles of evolutionary psychology one finds

> a triumph for Gould and Lewontin, who have seen almost all their original objections incorporated into the project. A nice example of this comes in Tooby and Cosmide's denial...that most of the differences between human beings are adaptive. (147)

He also shows that Wilson himself, in later publications, also introduced tremendous complexity into his analysis of even the mouse genome, let alone the human genome. According to A. Brown,

> this is an extraordinary retreat from the bright confidence of the original sociobiologists that everything could be reduced to the simplicities of genes and their kaleidoscopic patterns of adaptation. . . .

But then he goes on to conclude:

> Politically, however, the story seems to be one of steady marginalisation of Gouldians and the emergence and triumph of a refined and purified sociobiology. As a new orthodox hardens, the Gouldians' fate is clearly to be ushered into the history books as The Men Who Were Wrong. (147–148)

Likewise, Wilson's *Sociobiology: The New Synthesis*, which is to many critics one of the most ideologically laden formulations of human sociobiology, was reissued as a 25th-anniversary edition in 2000. Also, Dawkins' *The Selfish Gene* has been reissued as a 30th-anniversary edition (2006a). Despite the cogency and thoroughness of many of the critiques, it appears that sociobiology has flourished. As I indicated in the introduction to this book, the general project of associating biology and behavior has grown tremendously, and within academe sociobiologists, evolutionary psychologists, behavioral geneticists, behavioral ecologists, and others studying the connections between behavior, genes and evolution have created a consistent, funded, published, and "credible" paradigm. In addition, the more determinist formulation of the paradigm is still strong in the fields of behavior genetics, human behavioral ecology, and, to a certain extent, psychology, but not as significant in anthropological and sociological work.[1]

This chapter is divided into two large sections. In the first section are issues that are more internal to the structure of sociobiological discourse and that relate to larger questions within evolutionary theory and genetics. In the second section, the focus is on larger sociopolitical, economic, and ideological factors that have contributed to the popularity of sociobiological discourse. Throughout all of the sections and sub-sections in this chapter, I aim to articulate the many factors that have contributed to the legitimacy of a strong genetic paradigm especially within human sociobiological discourse, despite the many critiques aimed against it. And, again, as I indicated in the introduction to this book, these are more than just academic issues because new biotechnologies are beginning to make a consumer-based eugenics movement a possibility, and how we think about the influences of "nature" and "nurture" will inform the choices we make in the face of this emerging consumer eugenics environment.

Factors Internal to Sociobiological Discourse, Genetics, and Evolutionary Theory

While critics have attacked Wilsonian human sociobiology in many cogent ways and have had some effect on Wilson's own writings, they have been unable to permanently undermine the general credibility and legitimacy of the more determinist formulations of the project. When analyzing why and how it is that determinist conceptions of human sociobiology have been able to create legitimacy for themselves and grow in popularity both within academe and popular discourse, it is important to consider both internal and external factors. Throughout this work, I again remind the reader that, to indicate that I am not talking about just Wilson's work, I use the term "sociobiological discourse" to discuss conceptualizations that begin or take inspiration from Wilson's organizing texts but have proliferated now into many forms in many directions.

The Theoretical Context

Any knowledge generated through science practices, and specifically sociobiological knowledge, has strong social and political dimensions that affect the production, promotion, and success of Wilsonian human sociobiological discourse, and indeed all science claims. Along with this political and social dimension, the production of science knowledge also involves contestation and attempts by some to control the boundaries of the debate. According to Epstein,

> credibility describes the capacity of claims-makers to enroll supporters behind their arguments, legitimate those arguments as authoritative knowledge, and present themselves as the sort of people who can voice the truth. Credibility is, of course, a quality that can be established in many different ways in different arenas. The credibility of a speaker can rest on academic degrees, "anointment" by the media, or a speaker's access to esoteric forms of communication; the credibility of any knowledge claim can depend on who advances it, how plausible it seems, or what sort of experimental evidence is invoked to support it. (Epstein, 1996, 3)

It is therefore useful to analyze how Wilson and others who developed the sociobiology paradigm attempted to structure the legitimacy of their exercise and also how external forces have contributed to the popularity and legitimacy of human sociobiology. This work cannot constitute a complete treatment of all the ways in which Wilson and other "Wilsonians" construct legitimacy for their work. I will attempt to highlight those that I feel are the most significant.

The question of how scientists construct acceptance and legitimacy for their positions has been one of the main foci of the field of science studies for the past three decades (Shapin, 1995; Zuckerman, 1988). In his book *The Structure of Scientific Revolutions*, published in 1962, Kuhn showed that changes in scientific paradigms proceeded in a revolutionary manner involving contestation and controversy. He argued that, contrary to established positivist conceptions, these contestations and the ultimate acceptance or rejection of paradigms involved both disputes over evidence and the influences of social factors. Ever since Kuhn, philosophers, historians, sociologists, and scientists who study science have labored to articulate these social forces and their effects on theory and paradigm acceptance and legitimacy. I would place this work within the camp of the sociology of science. This field was begun by Merton (1970; 1973) in the middle decades of the twentieth century by examining the social influences on the development and proliferation of science theory.

Science as Discourse

I have written previously on nonsociobiological determinist discourses (Schifellite, 1987) and on gendered discourse in virology texts (Haddad and Schifellite, 1993). My analysis is both Kuhnian and sociological. That is, I see understanding the social influences on science as crucial to developing evaluations of science claims. My conceptual framework is influenced by a constructionist paradigm, in that I see the activity of scientists and the production of science knowledge as a social product. This knowledge is not neutral and is always socially mediated. This paradigm recognizes the important social character of knowledge and the importance concepts have on shaping our ideas about ourselves and about the world (Marx and Engels, 1976). My perspective on science knowledge is also influenced by Marxian theory, in that I see this knowledge as likely to have self-interest and social relations of power, inequality, and dominance embedded within it, and its content likely will reflect these influences back to society. According to Restivo,

> scientific and mathematical knowledge are not things "out there" in an eternal, universal Platonic realm that can be "discovered" in one revelatory way or another. Neither are they products of "pure" mental activity or of "geniuses" who create them out of thin air. In any given social formation, the prevailing mode of knowing grows out of practical activities (it is literally manufactured)....It reflects mediations among the various activities and products in a social formation. (Restivo, 1995, 101)

As a social product, science must be treated as a set of ideas that emerge as discourses or discursive practices that include not just ideas but individual and institutional practices and social relations that shape hiring, promotion, publication, and research funding decisions. This entire ensemble makes up what we call "science." Texts and discourses have a "material reality," or what D. Smith calls a "documentary reality." According to D. Smith, "Our knowledge of contemporary society is to a large extent mediated to us by texts of various kinds" (D. Smith, 1990, 61). D. Smith, echoing Marx's notion that ideas are material and social forces, develops the concept of "ruling relations," through which she describes the power and scope of this dynamic as it proceeds:

> In a lived world in which both theory and practice go on, theory is itself a practice, in time, and in which the divide between the two can itself be brought under examination. The entry into text-mediated discourse and the relations to text-mediated discourse are themselves actual as activities and the ordering of activities....Concepts, beliefs, ideas, knowledge, and so on (what Marxists know as consciousness) are included in this ontology of the social...as practices that are integral to the concerting and coordinating of people's activities....The social *happens*; included in the happening/activities are concepts, ideologies, theories, ideas, and so forth. (D. Smith, 1999, 75)

Latour and Woolgar (1979) and Latour (1987, 1988) also see the interplay of knowledge, power, and politics in the development of science knowledge. Epstein describes the process in the following way:

> . . . [S]cience is "politics by other means," and the credibility of a knowledge claim depends on the play of power: the scientist who can appear to make nature "behave" in the laboratory, whose rhetoric is more persuasive, who is able to summon up the more compelling citations, and who is able to enlist more allies, patrons, and supporters by "translating" their interests so that they correspond with the scientist's own is the one who constructs credible knowledge and gains access to further "obligatory passage points": the journal article that all must cite to justify their own work, the technology that all must employ to accomplish their own research—in general, the way stations through which other scientists must pass in order to satisfy their interests or achieve their goals. The more well traveled such passage points become, the more fully institutionalized the knowledge claims become.[2] (Epstein, 1996, 14–15)

I also am influenced by the work of feminist standpoint theorists such as Harding (1986, 1991, 1998), Longino (1990, 2000), and H. Rose (1994a) and by the work of theorists who have focused more on ideology studies, media studies, and on textual analysis and discourse studies. These theorists have done work on the ways that discourses are constructed and presented as legitimate. In this analysis of the structure and legitimacy of sociobiological discourse, I rely on the work of D. Smith (1987, 1990, 1999); Herman and Chomsky (1988); and Cohen (1989), among others. These influences will become more apparent as this chapter proceeds.

Science as Contested Discourse

There are a number of theorists—some who are not considered part of "science studies"—whose work I find useful in analyzing this discourse. In the examination of science as contested discourse, I draw from the work of Marx, Kuhn, Latour, and Gramsci. Kuhn and Latour see this dynamic of contestation in the development of rival theories and discourses in scientific disciplines. Latour sees that key scientists can play important roles in the propagation of paradigms as, for example, he sees Pasteur imposing his will and power on a scientific discipline. I do not pose a direct correspondence between the work of Marx and Gramsci and this work in science studies. Marx and Gramsci focused on the arenas of politics and class struggle. Gramsci constructed his theories about how hegemony shaped the process of ruling in a large political sense. For Gramsci, (1971) hegemony is the process of taking the specific interests and programs of a dominant group or class and formulating these interests and projects in ways that make them appear to represent the general interests of all groups or classes in a society. To do this is to build hegemony through leadership and consensus rather than relying on brute force to impose the will and interests of the dominant class or group. Gramsci was trying to understand large political movements and class politics, and to this extent his work is not about science studies. These areas are not the terrain of this work. I do believe, however, that we can use these general methods of analyzing the construction of ideas as this construction can intersect with positions of interest held by both groups and individuals.

Gramsci's (1971) notion of hegemony as a description of the processes wherein groups vie for dominance of their world views, or in this case, scientific views (the two often are closely related), is a useful mode of analysis that I employ at various points in this work.[3] Marx and Engels (1976) began the study of ideas as "material forces." Gramsci's work carried the examination of systems of ideas and social discourses into a new terrain. In the battle over "hearts and minds" within the sociobiology debate, both sides engage in strategies to legitimize their own positions and undermine their opponents' positions. Gramsci's work also is important in that it implicates consent, leadership, and transformation as important dynamics in the process of creating hegemonic discourses. The construction of hegemony is about controlling the parameters of debate. Gramsci's ideas have applicability when analyzing the construction of legitimacy of dominant scientific discourses as well. The general structure of his arguments can be applied to the sociobiology debate in a way analogous to the processes of contestation and legitimacy-construction that Latour and Woolgar found in their examination of *Laboratory Life* (1979).

The Hegemonic Aims of Wilsonian Discourse

This theoretical framework has relevance for analyzing the ways in which proponents of sociobiological discourse have attempted to build a hegemonic position. The processes that Gramsci (1971) articulated have applicability for the ways in which individuals or groups doing science build legitimacy, consensus, power, influence, and rewards—of whatever forms. Wilson, Dawkins, Barash, Dennett, Hamilton, Trivers, and Alexander have structured, and now others continue to structure, their claims so as to foster legitimacy, credibility, and popularity, and at the same time insulate these claims from critique. One can observe that during the sociobiology debate, the promotion of sociobiological discourse was not just about attempts at controlling discourses within seminal texts; it was also about control of things such as journals, research money, and faculty hiring and promotion. Likewise, it is about the ways in which larger social discourses and dynamics of power and inequality infiltrate and influence scientific claims and debates.

The original formulations of sociobiology appeared to be so sweeping, so aimed at reasserting the primacy of genetic influences, and at the same time carrying such political controversy and interest, that Wilson and others clearly became engaged in a hegemonic battle. This battle was, and continues to be, over the terms and boundaries of what can count as legitimate work or discourse involving theorization about the causes of complex human behaviors. It has been a battle over our understandings of broad concepts such as "traits," evolutionary theory, genetic explanation, and the nature of organisms. These emphases emerge partly because Wilson framed his work in broadly paradigmatic, sweeping, and controversial fashion, and partly because the nature/nurture debate itself has been so polarized and so hotly contested for at least a century and a half. In the post-World War II period, an extreme environmentalism emerged that attempted to limit the bounds of thinkable thought such that genetic arguments could not be conceived or considered. Wilson, Barash, Dawkins, and others have surely attempted to "bend the stick" in the other direction, and, if not place social analysis outside the realm of the possible, then at least to render that analysis secondary to the biological.

Wilsonian and Dawkinsian sociobiology is a coherent discourse framing what can be asked and answered in the contemplation of the "causes" of complex behaviors, much the same way that E. Keller sees twentieth-century genetics as discourses of "gene action" and "genetic programs" (E. Keller, 1995, 2000). In this regard she borrows from Hacking's idea that each discipline in science has a "style of reasoning" that frames the limits and contents of debate (Hacking, 1982, 1992). According to Keller (1995), these discourses frame the questions asked and control the organisms chosen, the experiments done, and the explanations that are acceptable (11). I would add to this the idea of contestation. In the process of D. Smith's "social happening," there also is likely to be contestation of what will count as legitimate descriptions of reality, as all knowledge reflects social influences and social realities. It is important to view these scientific paradigms not as monolithic structures but as sets of ideas, practices, and social relations that are sometimes dominant and sometimes subject to challenge and reinterpretation. This play of forces over time can be seen in the representations in the biology textbooks that I analyze in Part II of this book.

Constructing Hegemonic Leadership and Consensus

Wilson's project with *Sociobiology: The New Synthesis* was to construct both leadership and consensus for his "reading" of the theoretical developments that had been building in the 1950s and 1960s. In this creation of leadership and consensus, Wilson engaged in an expansive project that could encompass many disciplines. Segerstråle describes the creation of sociobiology "as the outcome of a *collective process*" [author's emphasis] (Segerstråle, 2000a, 85) in which she describes Wilson as having "created a field by demonstrating to its members that it existed—partly by co-opting them as contributors to his project" (314). And, according to Ruse (1999), Wilson engaged in grand theory building and

> whole new ways of looking at things—ways which hint at, if not promise, all sorts of fresh and exciting avenues of research. In a sense, the whole sociobiological synthesis is intended as an advertisement for new directions of scientific activity. (186–187)

Indeed, sociobiology did open up new avenues of research. Wilson's *Sociobiology: The New Synthesis* spawned a paradigmatic shift, and in doing so spurred the development of new lines of academic "disciplines." This development has in turn offered the possibility of generating many career-building ideas and offshoots. It is important not to underestimate the significance of the raft of new journals, societies, and research and funding categories, especially for new academics who must establish themselves by getting funded and published as often as possible. This fecundity has a self-referential and self-sustaining dynamic as it builds and also has contributed a great deal to bolstering the credibility and the dominance of neurogenetic explanations for human behavior.

Analogies, Metaphors, and Representations of the Specific as the General

Hand in hand with this process of building alliances is the process through which Wilson's synthesis elevates specific facts and theory to the level of general applicability. I am talking here about Wilson's attempt at creating a common calculus for understanding all behavior—both human and nonhuman—that I discussed in detail in Chapter 1. For example, it is perfectly reasonable to assume that much of what Wilson lays out in *Sociobiology: The New Synthesis* about genes, evolution, kin selection, and altruism can be used to accurately describe the factors and forces that strongly influence social behavior in insect societies. This does not mean, however, that these same principles can easily be extended to explain complex social relations and behaviors in higher mammals, let alone in humans.

For example, fewer critics challenged the need for a determinist formulation of behavior when Wilson discussed insect, avian, and even some mammalian social behaviors. E. Allen et al. (Sociobiology Study Group of Sciences for the People, 1978), however, took exception to what they perceived as Wilson's importing this model to human behavior: "While evolutionary analysis provides a model for interpreting animal behavior, it does not establish any logical connection between behavior patterns in nonhuman and human societies" (262). While it might be conceivable that there are similar behaviors and etiologies between primate and human behaviors, E. Allen et al. are critical of what they saw as Wilson's unsubstantiated movements from what might be conceivable "to what is" (262). Nonetheless, Wilson and others proceed as if that is the case. Indeed this was always Hamilton's project as well—to create a mathematics of kin selection that could be applicable to all creatures

(Segerstråle, 2000a). The movement from nonhuman sociobiology to human sociobiology is also an attempt to transfer legitimacy generally accorded the former, onto the latter—human sociobiology.

Wilson bolsters his generalizing with a generous sprinkling of anthropomorphisms. Critics from the beginning have held that Wilson's arguments in favor of a human sociobiology are fraught with anthropomorphism. For example, Kitcher (1985) cites Barash's and Krebs' and Davies' "use of 'rape' to cover behavior in scorpion flies, mallards and humans" (185). He also cites other instances including Barash's use of "talk of prostitution" in tropical hummingbirds and Dawkins' and Wilson's suggestion that "coyness" is likely to be a trait of "the courted sex" (189). He also contends that in addition to these more blatant examples, "there are more subtle cases, in which the choice of word allows the reader to draw unwarranted conclusions" (189). Also, Wilson uses terms such as "slavery" and "caste" to describe the social world of insects.

Likewise, Miller (1978), in pointing out the similarities between Spencer's and Wilson's lines of argument, holds that as with Spencer before him, "Wilson relies heavily on analogy and metaphor" taken from both things such as population biology and "metaphors from human societies" that are applied to animal societies (276). This is similar to the anthropomorphism I spoke of above. According to Miller, this kind of argument has two effects: "the hazy distinction between analogy and homology of the two types of societies becomes even more blurred; and metaphors tend to subtly corroborate the universality of such social institutions as slavery in the natural world" (276).

Miller (1978) also points out that Spencer and Wilson couched their science in the vernacular and in the common-sense perceptions of their times with the result that glib, superficial, and heavily ideological social analysis is given an evolutionary basis (276). This elevation of what may be proximally accurate in specific cases to general principles applicable in all cases is not just a theoretical act, it is also a political act because it at once establishes allies and circumscribes a hegemonic or paradigmatic domain for the entire sociobiology project.

Controlling Boundaries and the Limits of the Imaginable

In addition to providing leadership and the appearance of representing general interests, the creation of credibility and legitimacy within science as with other discourses also involves the ways that proponents and opponents attempt to control the boundaries of the imaginable. Gramsci (1971), Herman and Chomsky (1988), and D. Smith (1990) all see the establishing and controlling of the boundaries of debate and ultimately "thinkable thought" as crucial in maintaining control of a discourse and ultimately control of what D. Smith calls the "relations of ruling." This kind of "boundary work" is clearly operating in presentations of human sociobiology. Wilson "borrows" a great deal of legitimacy from other related disciplines such as genetics and evolutionary theory. To be fair, this is something that all scientists must do in order to conduct research and to make claims. This is the process (referenced above) that Latour describes in his discussion on "obligatory passage points" in the construction of scientific legitimacy.

Possibly some of the brilliance of Wilson's *Sociobiology: The New Synthesis* stems from the ways in which he has been able to create a paradigm in which so many key issues and social trends intersect, while encapsulating and referencing issues in science, philosophy, human nature, genetics, and evolutionary theory in particular ways. Sociobiological discourse has attempted to stake out particular interpretations of key concepts in genetics, evolutionary theory, and epistemology. In Chapter 1, I spoke about the dissatisfaction of critics with sociobiology's use of a simple Mendelianism to describe the ways in which genotype and phenotype are related. Ultimately, this disagreement references a much larger set of issues within genetics that relates to our understanding about the functioning of genes, the conceptualization of traits, and the nature of organisms. As is evident in subsequent subsections in this chapter, the positions that sociobiological discourses borrow from genetics and evolutionary theory are not the only positions available. Also, what critics find is that the debate quickly moves away from the debates within sociobiological discourse to disagreements about core ideas in genetics and evolutionary theory. In all of these cases, there are alternative positions and interpretations in the home disciplines that either must be left unacknowledged or marginalized. Part of the process of creating hegemony involves attempting to support one set of interpretations while undermining the credibility of the positions that support critical positions. Ultimately, the sociobiology debate moves from critique to the formulation of alternative and possibly competing discourses.

The Use of Language in Sociobiological Discourse

The Language of "Lumbering Robots"

In its many popularizations, sociobiological discourse has played fast and loose with language and concepts that connect genes to behavior. This is an arena in which sociobiological discourse attempts to fix the boundaries of what is conceivable and to have genetic programs control organisms as if they were "lumbering robots (Dawkins, 1976)." Kaye (1986) sees this portrayal as, to some extent, the accidental outcome of the looseness of popularizations. According to Kaye (1986),

> *The Selfish Gene*, Dawkins tells us, "is designed to appeal to the imagination" and to be "astonishing," "entertaining," and "gripping" in order to convert readers to the new "truth" that human beings are really "survival machines—robot vehicles blindly programmed to preserve the selfish molecules known as genes." Barash's *Sociobiology and Behavior*, although designed as an introductory scientific text for college students, is admittedly "one-sided" and "speculative," a "persuasive primer" advocating the sociobiological approach to all social behavior. (136–137)

[and]

> . . . both Dawkins, author of the widely read *The Selfish Gene* (1976), and David Barash, an American zoologist and author of *Sociobiology and Behavior* (1977) and *The Whisperings Within* (1979), recognize that to speak of "selfish genes" and the reproductive "choices" and "strategies" of genes and organisms is "sloppy" and misleading. They claim that such "anthropomorphisms" are used solely to simplify and render "palatable" what are difficult scientific ideas. . . . (136–137)

A. Brown sees a similar conceptual loading having happened with Hamilton's use of the metaphor of intentional or intelligent genes. A. Brown shows that in Hamilton's (1964) seminal paper entitled "The Genetical Evolution of Social Behavior," Hamilton uses the idea of genes with agency and intelligence only as an explanatory metaphor (A. Brown, 1999, 23). Hamilton uses this metaphor of gene intention to "try to make the argument more vivid by attributing to genes, temporarily, intelligence and a certain freedom of choice" (cited in A. Brown, 1999, 24). In so doing, he encourages the reader to "imagine that a gene is considering the problem of . . ." (cited in A. Brown, 1999, 24). A. Brown points out that after elucidating the point he is trying to make,

Hamilton then reminds the reader: "We can now abandon the fanciful viewpoint of individual genes" (cited in A. Brown, 1999, 24). A. Brown points out with irony, however, that "seldom has so clever a man been so wrong. [His] mathematics are remembered by a few professionals, [that] metaphor has run around the world" (24).

Wilson, Hamilton, and Dawkins are surely partly responsible for giving this metaphor a life of its own. In the first page of *Sociobiology: The New Synthesis*, Wilson makes it clear that he sees organisms as not existing for themselves but simply having the primary function of reproducing genes: "the organism is only DNA's way of making more DNA" (1975, 4). Dawkins (1976) carries this perspective further by transferring any notion of intention from the "lumbering robots" into the genes themselves. Instead of animals "deceiving," "trying," or "investing," genes become "selfish"—this work of intentionality and intelligence is given over to genes themselves.

Wilson, Dawkins, and many defenders of sociobiology have long held that they use reduction only as a conceptual tool and that they use metaphors such as gene intentionality as explanatory vehicles. Indeed, A. Brown (1999) makes the argument that Dawkins is a sophisticated thinker and not a simplistic reductionist or determinist. He holds that the disagreements between opposing sides are, in reality, small and that the real disagreements "about what genes are and how they affect the world...are subtle" (30). This sentiment is echoed by Segerstråle (2000a).

Nonetheless, A. Brown also contends that when you ask what Dawkins really meant by selfish genes and how deterministic his ideas really are, "you may well come away confused," and for Brown, this confusion "is not unimportant or even, in a sense, accidental" (1999, 30). Kitcher (1985) is harsh on this general point. For him, "Verbal tricks abound in pop sociobiology, serving as substitutes for argument or vehicles for misleading suggestion" (190).

Rhetorical Devices: "Mentioning," Transformism, Doublespeak, and Misdirection

Sociobiological discourse also relies on a number of what I would call rhetorical devices to bolster its legitimacy. These include the devices of "mentioning," doublespeak[4] and the "propaganda of the middle ground." Before going on to discuss these, however, I will briefly describe the Gramscian notion of "transformism" and its relation to this discussion. Gramsci developed the concept of transformism to describe the way in which a class or group can maintain its hegemony in the face of critics or protests (Gramsci, 1971, 58). In this process, the dominant group is able, through the state, to transform and adapt criticism directed against it so as to incorporate these attacks while not sacrificing real power, domination, and control. It is a form of acknowledgment that can build consensus but ultimately maintain power and control.

Wilson and others have engaged in an analogous rhetorical process. They have attempted to acknowledge and deflect the critics' objections and insistence on including cultural, cognitive, existential, and socioeconomic variables in an analysis of human behavior. So there is a kind of lip service being paid to these critiques and to the issues raised in the literature while they propose no real theorizing of the complexity of such interactions and no undermining of the ultimate biological control of behavior. Doloris Durkin, in the context of textbook analysis has called this process "mentioning" (cited in Tyson-Bernstein, 1988, 27). Tyson-Bernstein describes the term in the following way: "Books accused of 'mentioning' are generally long on facts and terms but short on ideas and explanations" (27).

Tyson-Bernstein (1988) has observed that when textbooks treat topics superficially "students can't make sense of what they are reading" (19). According to Tyson-Bernstein, the rational reaction by students to topics that are "mentioned" is "so what?" or "who cares?" This concept is also discussed by Apple and Christian-Smith (1991). These authors give it an interpretation that is more analogous to Gramsci's notion of transformism. They see textbook authors as attempting to mollify those critical of the hegemonic function that seems built into school texts by "mentioning" specific issues. This act of "mentioning" apparently addresses the issues raised by critics without really doing so (10). This kind of mentioning clearly is employed in texts that promote the Wilsonian sociobiological discourse. It also is apparent in some of the presentations of sociobiology in the textbooks that I present in Part II of this book.

"Mentioning," especially as practiced in Wilsonian human sociobiological discourse, can lead to a confusing doublespeak. Sometimes Wilson seemed to advocate a strong determinism and at other times, especially in the face of criticism, he advocated a more nuanced position. This back-and-forth movement in Wilson's work and in similar kinds of work that has followed Wilson's, has tended to lead critics to see sociobiology as fundamentally determinist or, at the very least, misleading and confused. This pattern of doublespeaking seems clear in the following passage of Wilson's (1978):

> . . . *general sociobiology allows three alternative states for our species:*
>
> 1. Natural selection has exhausted the genetic variability underlying social behavior: human populations are uniform with respect to the social genotype. Furthermore, the genotype prescribes only the capacity for culture; in this sense human sociality has been freed from the genes,
> 2. The social genotype is uniform, but prescribes a substantial amount of instinct-like behavior.
> 3. Some variability in human social behavior has a genetic basis, and, as a consequence, at least some behavior is genetically constrained. (xi)

E. Allen et al. (1978) argue that one can see this in Wilson's (1975) discussion of what he calls Theodosius Dobzhansky's "extreme orthodox view of environmentalism" that Wilson summarizes as: "In a sense human genes have surrendered their primacy in human evolution...to culture." The authors then note that Wilson ends this discussion by reversing this position and saying "the exact opposite could be true" (550). And, "suddenly, in the next sentence, the opposite does become true as Wilson calls for the necessity of anthropological genetics" (550).

At one level, all of this backtracking and hedging is understandable and speaks to the complexities involved in trying to theorize the connections in humans between genes, hormones, brain chemistry, cognition, culture, and complex behaviors. At another level, however, this backtracking and hedging seems to serve at the very least to keep the reader off balance and confused about what is being claimed, while, at worst, allowing Wilson the chance to make strongly determinist arguments at will. Doublespeak, more than anything, confuses especially naive or lay readers and deflects justified critique.

In conjunction with Wilson's attempt to displace, transform, or co-opt criticism when possible, I also see him invoking a strategy of claiming the balanced "middle ground" between supposed extreme positions in a way that is reminiscent of what Cohen (1989) calls the "propaganda of the middle of the road." They use this phrase to describe the ways in which debate within the major news media is constrained. He argues that there is an implicit assumption, fostered by the dominant media, that there is propaganda of the right and left, while the middle ground—that incidentally is occupied by the media—has no propaganda or bias.

This rendering of the dominant media coverage as "the middle ground" serves to hide their biases and to create a patina of neutrality and credibility for their positions. The use of doublespeak, anthropomorphisms, generalities, and the staking out of the "reasonable" middle ground are all rhetorical devices. They all work to foster the impression of reasonableness and balance, and they can be effective in deflecting both political and theoretical critiques of their work.

Legitimacy Through Selective Interpretation in Evolutionary and Genetic Theory

Ultra-Adaptationism

As indicated in the previous chapter, one of the key boundary areas within sociobiological discourse involves disputes about core ideas in evolutionary theory. Gould feels that sociobiological discourse supports a strong or ultra-adaptationist position that is not shared by all who work on evolutionary theory. According to Ruse, Gould feels that an overwhelmingly adaptationist Darwinism "is not wrong, but it is a very limited part of the picture" (Ruse, 1999, 142). According to Segerstråle, Gould's strong objection—on anti-adaptationist grounds—to Wilson's formulations in *Sociobiology: The New Synthesis* has been that he feels that prior to the 1950s a kind of pluralism held sway within evolutionary theory. This pluralism gave space to adaptationism and other kinds of evolutionary arguments but "was temporarily suppressed when in the 1950s the Modern Synthesis 'hardened' into a belief in the pervasive power of adaptation through natural selection" (Segerstråle, 2000a, 111).

This focus on adaptation is also related to a conception of natural selection that transforms evolutionary theory into a teleology in which 'that which is, is all that could have emerged.' This teleology is not something new to sociobiology. Etkin observed that natural selection "is itself pseudoteleological" (Etkin, 1981, 50). That is, it lends itself to the use of purposeful language in discussions of the workings of selection. Selection operates, however,

> not in terms of whether they lead to an adaptive goal...but rather of whether they have immediate competitive advantage over alternatives....[thus] The terminological confusion arises from the fact that the long-term effect on the species—that is, the evo-

> lutionary outcome—is most easily described in the teleological thinking (anthropomorphisms). (50)

He claims that biologists today understand that they are using a heuristic device when they employ this kind of language. Nonetheless, he warns that there is a danger that this language can "sometimes obscure the scientific interpretation, confusing the author's analytical thought with the animals 'motivation'" (50). This can be especially true for works of popular science in which the audience may be unaware of the implicit distinctions and qualifications an author employs.

Historian Robert Young goes a step further. Darwin was writing before modern genetics and the concept of the gene, so it was the concept of natural selection that became loaded with ambiguity and intention. Young has long held that this kind of teleology was a fundamental part of Darwin's own formulations of the concept of natural selection. Young sees Darwin as having been influenced by both his study of the conscious work of animal breeders and also by the omnipresent natural theology of his time that saw a divine hand behind the order in nature. For Young, the concept of natural selection from the outset "was anthropomorphic, deeply ambiguous and amenable to all sorts of readings and modifications" (Young, 1985, 125).

Gould saw resistance to these hard-line adaptationist and selectionist positions emerge throughout the 1960s and 1970s and holds that Wilson's *Sociobiology: The New Synthesis* "reasserted the Panglossian form of adaptationism just at a time when much of the excitement in evolutionary theory was breeding departure from this former orthodoxy" (cited in Segerstråle, 2000a, 111). In this way, one element of the sociobiological debate has been about what will count as the "correct" emphasis for the role of adaptation in evolution and what will count as the "correct" understanding of the process itself. Critics attempting to undermine the legitimacy of a strongly genetic sociobiological discourse quickly find that a critique of a strong adaptationist position moves them, in effect, outside the boundaries of the immediate discourse into a much larger debate about what should count as legitimate concepts in evolutionary theory. So long as this remains unsettled for evolutionary theory, adaptationist conceptions will be employed within strongly determinist formulations.

Simple Mendelianism vs. the Complexity of Organisms

Critics have been forced to wage the battle against reductionism in sociobiology, not just in the arena of evolutionary theory but also in genetics. With the discovery of the structure of DNA and the establishment of the modern synthesis, genetics became, along with Darwinism, the dominant paradigm in biology. In recent decades, genetics has become more and more dominated by molecular genetics. During this entire process there has been a background, and sometimes foreground, debate about the connection between genotype and phenotype and between the gene and the traits supposedly under these genes' control. This debate has been especially heated when human beings are considered and the history of genetics is tainted by the actions of leading geneticists in Britain and the U.S. who advocated and succeeded in implementing forced sterilization campaigns in the name of eugenics in the first half of the twentieth century (Kevles and Hood, 1992). As I indicated in the last chapter, after World War II, geneticists such as Dobzhansky advocated the position that the causes of human behavior and human culture could not be considered to be genetic and that separation held for a few decades.

Wilson, Dawkins, and others in human sociobiology clearly side with the very strong program of genetic determinism that has existed and still exists within genetics today. As Hubbard (1982) argues, however: "[N]othing in the concept of factor, gene, or genotype implies a causal line to characters or traits" (65). She argues that genes produce only proteins and that Mendel's simple paradigm works only for a small fraction of the patterns of inheritance of mutant genes, and "most traits do not follow Mendel's Laws" (71–72). While geneticists are well aware of the complexities of which Hubbard speaks, she argues that there is still an underlying idea that genes at a fundamental level are in control of organisms. This principle is evident in the central dogma of molecular genetics of which she is critical and that holds "that *information* [author's emphasis] passes in only one direction from DNA to RNA to protein" (69). In *Exploding the Gene Myth*, she and Hubbard and Wald (1993) added:

> The language that geneticists use often carries considerable ideological baggage. Molecular biologists, as well as the press use verbs like "control," "program," or "determine" when speaking about what genes or DNA do. These are all inappropriate because they assign far too active a role to DNA. The fact is that DNA doesn't "do" anything; it is a remarkably inert molecule. It just sits in our cells and waits for other molecules to interact with it. (11)

A bind for critics is that they must either hold sociobiology up as bad science for the simplistic way it presents issues in genetics, or they must critique genetics itself for employing simplistic models. This shifts the debate away from sociobiological discourse and into the heart of the canon of genetics. Critics of a strongly determinist sociobiological discourse have argued that, within genetics, more sophisticated models concerning the relationship of genes to behavior are used. While this is true, the problem is that both kinds of formulations are in use within genetics. If critics take this route, then defenders of the discourse can point to places in genetics from which they borrow their formulations. This leaves the critic trying to clarify or reformulate ideas in genetics so that they conform to the underlying positions the critics hold concerning the nature of the object of study and the nature of science itself. Ruse's (1978) early defense of Trivers exemplifies this process:

> ... [T]he critics object that when someone like Trivers wants to explain human altruism on his model of reciprocal altruism, the whole enterprise is fundamentally misconceived and muddled because it is just not the case that there are, or even could be, genes corresponding to man's various altruistic traits. The genes just do not work that way....[W]hile it is certainly not the case that organisms can be divided neatly into an "objective" set of characteristics, or that there is a perfect match between characteristics and the genes, there is at least some correlation. To argue otherwise is virtually to reject genetics as presently conceived! (363–364)

In *Lifelines*, Rose (1997) seems to be arguing the opposite position when he says that "to biochemists, if not geneticists, there is no longer any gene 'for' eye color" (115). There is an interesting ambiguity in this statement. It holds that biochemists understand, but some geneticists may not. These are core issues and they are controversial and cannot result in easy resolutions. It is important to understand that some of the legitimacy and support for the Wilsonian human sociobiological project exists because many in genetics appear to believe or at least act as if they assume that behaviors can be understood as traits and that a simple Mendelian model can be used to explain and predict these behaviors. Simply criticizing Wilson and others for employing a model that has support in genetics will not necessarily undermine the legitimacy of the model inside genetics. The debate around these controversial issues must be shifted away from sociobiology and into genetics and evolutionary biology themselves. The ambiguities and differences existing within genetics and evolutionary biology must be articulated and challenged as part of a larger project to counteract a particular kind of biologization and geneticization of the life and social sciences.

Challenges to reductionist paradigms have been ongoing even as reductionism in biology has regained its footings in genetics and evolutionary biology. Dobzhansky led this challenge in the post-World War II years. One of his students, Ernst Mayr (1982), in the very thorough work *The Growth of Biological*

Thought, continued critiquing this reductionism. G. Allen (1983), in reviewing this work, points out that Mayr's critique of reductionism is not a simple vitalism that creates a privileged place for biology (25). Rather, he argues that Mayr understands the complexities involved in the nature of nature.

> While Mayr makes it clear that he does not argue that biology defies the laws of physics and chemistry, he clearly shows that it cannot be reduced to them....Mayr argues that the natural world must be viewed hierarchically. More complex systems derive out of, but are more than mere extensions of, less complex systems. They have emergent qualities—not mystical qualities....[They have] what the dialectical materialist would recognize as qualitative differences arising out of quantitative differences. (G. Allen, 1983, 25–26)

Allen also goes on to cite a number of scientists who have challenged "the mechanistic outlook in biology" for many years (29). He includes in this list "Robert Young, Sylvan Schweber, Stephen Jay Gould, Richard Lewontin, Steven Rose, Bernard Norton, Donald Worster, John Greene, and Sandra Herbert, among others" (29).

One of the more interesting and groundbreaking pieces was written by Ruth Hubbard, in an important essay entitled "The Theory and Practice of Genetic Reductionism: From Mendel's Laws to Genetic Engineering" (Hubbard, 1982). In this paper, Hubbard gives an interesting account of how difficult a project it is for her, as a trained scientist, to develop critiques of reductionism.

> This is the first time I have written these ideas out in a systematic way. In formulating them, I find it a disadvantage to have been trained as a biochemist; for though this training makes it very easy for me to think in molecular terms, it also makes me comfortable with the sort of molecular reductionism that I want to evaluate critically. To be suspicious of one way of thinking does not necessarily arm one against it—at least not all the time. (76n)

In this paper, Hubbard argues that

> genetics, the systematic description of hereditary mechanisms, to a large extent is a reading into nature of the twin ideologies of hereditarianism and individualism that were dominant during the period of its invention and proliferation. (63)

She also argues that a tension has existed within genetics for a long time

> between the interpretations of genes as intrapersonal determinants of characters and as causes of who and how we are, and the more modest and realistic claims based upon Mendel's Laws, that provides ways to analyze formally certain visible manifestations (characters, traits) that change (i.e., "mutate") in predictable ways in successive generations. (63)

And, in the post–World War II years,

> shifts of outlook brought into genetics ways of thinking about coding and control that are considerably more mechanistic, reductionistic, and control-oriented than was the cytogenetic and biochemical thinking that prevailed through the 1940s. (69–70)[5]

Hubbard goes on to outline the process by which concepts such as gene, phenotype, and genotype were created to flesh out the details of this picture of the way genetic material is reproduced and produces organisms.

Likewise, there have been historians who have begun to piece together the ways in which the current ideas in genetics have been structured and become dominant. Sapp (1987, 1990a) shows the ways in which Mendelianism and a focus on the cell nucleus as the controller of heredity becomes a dominant paradigm within genetics and this version of genetics within biology. Diane Paul (1998) has done work in *The Politics of Heredity* on the politics and history of the eugenics movement and the nature/nurture debate. Garland Allen (1983, 1994) also has done work on the history of the development of the early twentieth-century eugenics movement within biology. More work on the history and structure of argument within genetics is needed. While a great deal has been written on the history of genetics in the first half of the twentieth century and on the eugenics movement in particular, less has been written on the developments following World War II.

The Reification of "Traits" and the Nature of Organisms

Sociobiological discourse abounds with concepts such as selfish genes, genetic altruism, and adaptive homosexuality, but to what biological processes do these concepts actually refer? What behaviors can and should be considered traits? There is no clear answer to these questions, and sociobiologists have been able to exploit these ambiguities in the shaping of their arguments. Central to this presentation of organisms as robots controlled by genetic programs is an amorphous conceptualization about what can count as or what can be a biological trait. This is not really a problem that begins with sociobiology. Rose (1997) has pointed out, when tracing the historical development of concepts such as gene, allele, phenotype, and genotype, that it is important to note:

> none of these terms was very precisely defined, and almost from their introduction they meant different things to different researchers, varying from the specific features of any individual of a species to some Platonically idealized "species-type" to which all actually existing more or less approximated. (102)

He is pointing out a serious problem in genetics. Many of the key terms have a variety of sometimes contradictory meanings. This is especially true for the concept of "trait" or "character." This is an important point for critics of neurogenetic determinisms to consider, because it is once again a crucial site for boundaries disputes in the sociobiology debate.

The concept of trait or character is a keyword in the sociobiological arsenal. Its ambiguous meanings, within genetics, to some extent, allow Wilson and others to do "trait fixing"—whereby complex, diverse, and non-uniform social phenomena become "fixed" as biological, adaptive, Mendelian, and ultimately genetic and molecular entities.[6] It is necessary to challenge not only reductive formulations in genetics but also to challenge conceptual ambiguities that reinforce this reductionism. This has begun to happen. In *Keywords in Evolutionary Biology*, Darden (1992) briefly examines historical perspectives on the term "character" and concludes that it is "not an unproblematic, easily operationalized concept, even in contemporary biology" (44). Fristrup (1992), in the same book, analyzes current usages of the term "character" and concludes:

> The varied uses of "character" reflect the different goals of historical reconstruction and research in ecology and natural selection. Additionally, a source of confusion has been failure to distinguish between characters as parts, attributes, or variables; are characters natural units or artifacts of observation and description?...Some use character to refer to unprocessed observations; others introduce additional restrictions or analyses to produce characters that more closely resemble the information they would most like to have. (51)

Rose argues in *Lifelines* that many of the terms used in modern genetics and evolutionary theory today have multiple and often confused and confusing meanings. He further argues that it is imperative that scientists standardize the meanings of key terms. As long as conceptual vagueness remains, sociobiological discourse is free to create self-sustaining teleologies that biologize a selected behavior, call it a trait, and then plug it into a strong adaptationist paradigm and ultimately naturalize it as a fundamental part of human nature. To break this self-sustaining teleology, critics must begin to challenge, clarify, and circumscribe the definition of "trait" or "character" that is used in biology.

Social scientists have an important role to play in this process. D. Smith's work, extrapolated from Marx, is useful in examining this process of creating self-sustaining and self-referential ideological circles in the context of what she calls ideological practices in sociology. The process she outlines is very relevant to the study of human sociobiological arguments (D. Smith, 1990, 43–45). For Smith, in this process of creating what she calls "ideological circles" the actual activity of people engaged in interactions and social relations becomes reconceptualized as a noun—a thing that stands outside the human activity which gives rise to it. Further, she holds that "to treat assumptions about human nature (among other concepts) as active forces in social and historical processes is an ideological practice" (D. Smith, 1990, 36).

Taking a cue from Marx, she sees concepts as "a kind of 'currency'"—as entities that stand in between object and knower (D. Smith, 1990, 42). If we let concepts become divorced from the contexts in which they are created, our work becomes ideological.

> To think ideologically, by contrast, identifies methods of reasoning that confine us to a conceptual level divorced from its ground; we remain then "on the side of" the concept; the internal relations in the observable between concept and the actualities of co-ordered activities is ruptured. Concepts then become a boundary to inquiry rather than a beginning. (41)

In the case of sociobiology, what gets created are "traits" that then become recast again and abstracted again as "genes for traits." In this double abstraction, the abstracted concepts then stand in for the original activity and are seen to cause that activity. Volition is abstracted from people and organisms in general and relocated in genes. There is also a triple level of abstraction occurring because these abstracted concepts of traits and genes obliterate what is actually occurring at the molecular level—namely, the complex processes of protein synthesis dictated by these genetic sites. These processes of abstraction allow Wilson and others to engage in this movement from the specific to the general that I described in Chapter 2. In this movement, concepts that may be valid for insects or other less neurologically complex organisms get to stand in for people and their biological and social lives.

These abstractions also gain legitimacy from the larger set of social relations in society and reinforce those relations. For critics, explicating the connections between the social and the biological are crucial to exposing the ways in which larger social forces and this process of reification can appear to "naturalize" human behaviors and to naturalize constructed inequalities. Echoing D. Smith,

Rose notes that

> the resonance of "gay brains" or "selfish genes" does more than merely sell books for their scientific authors; it both reflects and endorses the modes of thought and explanation that constitute neurogenetic determinism, for it disarticulates the complex properties of individuals into isolated and localized lumps of biology. (S. Rose, 1997, 288–289)

Harding, using D. Smith's work as a basis, provides an example.

> [D. Smith] points out that if we start thinking from women's lives, we (anyone) can see that women are assigned the work that men do not want to do for themselves, especially the care of everyone's bodies....And they are assigned responsibility for the local places where those bodies exist as they clean and maintain their own and others' houses and workplaces....This kind of work, she shows, frees men in the ruling groups to immerse themselves in the work of abstract concepts. *The more successful women are at this concrete work, the more invisible it becomes to men as distinctly social labor. Caring for bodies and for the places in which bodies exist disappears into nature. Consider, for example, sociobiological claims about the naturalness of altruistic behavior and domestic work for females and the unnaturalness of either for males* [author's emphasis]. (Harding, 1998, 152)

What we observe as universal human behaviors are the products of social forces and social relations of organisms in interaction. These events are turned into abstractions and "fixed" as traits and ultimately genes, and these abstractions then replace actual activity and become the "causes" of that activity.[7] This simplistic presentation is dangerous, as Rose notes, because "thinking about genes as individual units that determine eye color may not matter too much, but how about when they become 'gay genes' or 'schizophrenic genes' or 'aggression genes'"? (S. Rose, 1997, 116)

This is not just a problem with the older texts by Wilson or Dawkins. The Wilsonian legacy continues today in many forums. One can see this process happening in the research that is ongoing in the investigation of the biological and genetic roots of male homosexuality. According to this research, correlative studies on sexual orientation have long seemed to indicate that there may be a genetic component involved in the development of male homosexuality and that expression of this "trait" is "quantitative, almost certainly involving multiple genes" (Clark and Grunstein, 2000, 242). In the early 1990s, Dean Hamer and his colleagues claimed to have located a genetic marker on the X chromosome. This marker, called Xq28, appears in an unusually high proportion of gay men. Rothman (1998) now reports that Xq28 has been "officially recorded as GAY1" (211). This has been done despite the fact that many have been critical of Hamer's work and a subsequent Canadian study "failed to find preferential inheritance of markers in this region in male homosexuals" (Clark and Grunstein, 2000, 246).

This is a clear example of this process of trait fixing. Male homosexuality as a set of practices and relations carried out by people with distinct social and biological histories has been transformed, reduced, and inscribed as one or more genetic markers on chromosomes. These markers then become seen as causes manipulating the people they inhabit. It may be that homosexuality, like schizophrenia and cancer, is not one thing but a variety of factors and events that reductionists have categorized and abstracted as one concept—one noun. By doing so, they obliterate what might be a complex and diverse field of biology and experience that resists categorization and prediction in traditional reductive and neopositivist ways. This process of reducing complex organisms and their social relations to abstracted concepts such as "genes" becomes similar to D. Smith's idea of ideological practices and Herman's and Chomsky's concept of controlling the boundaries of thinkable thought.

Critics must fight against the disappearance of the complexity of organisms, human and nonhuman, and the complexity of the social, political, and cultural spaces they inhabit. What is really at issue here is what can count as traits in both biological and social contexts. The movement from a complex social world directly to biological or genetic causality requires that one deny the importance of the social dimension. It requires that one treat complex behaviors, beliefs, practices, and institutions as monolithic and universal categories and that one take these complex interactions and reduce them to biological traits. Critics of this process must argue that the "facts" of social analysis such as behaviors, cultural practices, homosexuality, and even suicide are not simple facts out there waiting to be observed, but events that are constructed within and by social relations and practices. What counts as homosexuality? What counts as a suicide? These are designations that people give to themselves or are given to them through what Smith calls a process of inscription in documents, laws, social relations, and social practices.

McKinnon (2005) echoes this sentiment in what she sees as evolutionary psychologists' transgressions against the integrity of the cultural dimensions of human existence, especially as studied by anthropologists. According to McKinnon,

> at a time when there is an urgent need for a nuanced understanding of the complexities and varieties of social life, evolutionary psychologists provide instead astonishingly reductive myths and moral tales. (2)

One should not be surprised that Wilson and others tried to create hegemony for their positions. These scientists can fall prey to the same self-interested positioning as do the rest of us. One must, however, as McKinnon suggests, require that they subject their ideas to "normal" scientific scrutiny and to contrary evidence (2005, 11–13). They must not be allowed to rely on rhetorical devices and manipulations. Likewise, in order to debate the merits of each paradigm, we must understand not just the rhetorical devices used internally to support a position; we also must understand that there are external factors and larger ideological contexts that also conspire to resonate with and reinforce a hegemonic position for deterministic sociobiological discourse. Again, McKinnon echoes this idea when she says that what evolutionary psychologists "consider universal aspects of sex, gender and family are actually dominant European American conventions" (13). Sociobiological discourse and its descendants formulate theory that resonates with and reaffirms core dominant ideological beliefs.

External Forces in the Sociobiology Debate

There also are several external factors that help to build legitimacy, credibility, and popularity for determinist human sociobiology positions. All of these factors reference issues and forces external to the sociobiology debate itself, and critics can have limited influence on them. The first of these factors is the general advances and successes in genetic theory and bio-technological practice. Successes in these fields provide both legitimacy and a framework from which to advance all manner of genetically based causes for complex human behaviors. It is difficult to keep up with developments, to integrate this knowledge into non-determinist formulations that can shed some light on how biology influences behavior, and also to guard against the most extreme neurogenetic formulations. A second set of factors also is primarily external to the debates and beyond the reach of critics. These are market forces within the publishing industry, sociobiology's conceptual resonances with the larger set of dominant ideas in society, and foundation and governmental funding decisions and supports.[8]

I see the analysis of building legitimacy as a type of Gramscian analysis, and this kind of work generally calls for the crucial element of class analysis. In this case, such an analysis would mean examining the role of class in the promotion and propagation of Wilsonian sociobiological discourse. I do believe that such direct class-based connections may exist in the history of sociobiological discourse, and that one may find elites propagating sociobiology as a way of promoting their own continued dominance. In late capitalism, however, class influence over science knowledge has become, to a large extent, mediated through state and private funding and educational institutions. Finding the "smoking guns," in this case, would require an in-depth historical analysis, which is beyond the scope of this study. What I have done here, in this discussion of external influences on sociobiology, is briefly discuss, in a preliminary way, two arenas—foundation and state-sponsored support and publishing industry support—that have contributed to the rapid expansion of Wilsonian sociobiology. I also have looked at the ways in which the larger, dominant social and political ideologies interact and resonate with Wilsonian sociobiological discourse to the benefit of both camps.

Government and Foundation Funding Practices

The ways in which governments and private foundations allocate funds have had a significant effect on the development and acceptance of the general formulation of biological determinism and on the specific development of sociobiology. Garland Allen has done research on the individuals and foundations that supported the Eugenics Record Office at Cold Spring Harbor (G. Allen, 1991, 1994). Diane Paul (1991) has done work on "The Rockefeller Foundation and the Origins of Behavior Genetics" (Paul, 1991). Donna Haraway has done work on the effects that the Rockefeller Foundation funding decisions had on the decline of the Theoretical Biology Club of Cambridge, England, and its non-reductive formulations of development and evolution (Haraway, 1976). Lily Kay (1993) has chronicled the ways in which both the Rockefeller Foundation and the California Institute of Technology helped create a hegemonic position for molecular biology within the biological sciences. Also, Pnina Abir-Am (1982) has done work on the influence that the Rockefeller Foundation's policies had on the engineering of molecular biology's dominance. Much more work is needed in the history of the funding of neurogenetic determinist work. S. Rose points out the ways in which the Rockefeller Foundation "concentrated its resources on the sciences of psychobiology and heredity" in the early part of the twentieth century (Rose, 1997, 273). He also points out the ways in which "the sheer power and scale of the Rockefeller vision, backed as it was by hundreds of millions of dollars, ensured that alternative understandings of biology withered" (274).

In addition, Segerstråle has chronicled the crucial importance of the 1969 Man and Beast conference held in Washington, D.C., in May of that year. This conference was organized, lavishly funded, and hosted by the Smithsonian Institution, and she contends that it "may well have had a catalytic effect for development of sociobiology" (Segerstråle, 2000a, 90). In corroboration, she cites Wilson's observation that the conference "was an early milestone in the development of sociobiology" (91). It was very high profile, bringing together researchers from all over the world. According to Segerstråle (2000a), "Bill Hamilton, one of the invitees, describes it as a grandiose operation, involving lavish receptions...with speeches by Nobel laureates and eminent politicians (90).

There are two very interesting things about this conference. The first is the presence of so many dignitaries and eminent scientists. Such a gathering serves to convey a patina of legitimacy to the enterprise of connecting biology to behavior. The second is what Hamilton reports as an overarching concern among the prominent politicians in the crowd with social issues such as crime and violence in the destabilized period of social unrest of the late 1960s. His overall impression was "that the conference was called to invite biologists to formulate solutions to current social ills" (Segerstråle, 2000a, 91). These connections between foundation and government support and the development or withering of branches of scientific investigation are very important and deserve much more attention. They play a key role in the development of credibility and legitimacy for some science claims and a key role in the development of hegemonic positions for specific science paradigms and in the disappearance of others.

The role that these institutions have had and continue to have in fostering hegemonic or dominant positions for genetically and biologically determinist approaches to the study of human behavior is part of a process that I would call "boundary work." As I indicated in the previous section on internal factors, an important part of creating a set of ideas and practices that are dominant and that determine how people see a given piece of reality is the process of creating boundaries around what will be considered as acceptable fact and theory. Herman and Chomsky, in their work on propaganda, examine the importance of "limiting the boundaries of thinkable thought in the manufacture of consent" (Herman and Chomsky, 1988, 1–35). D. Smith (1990), in the development of the notion of "ideological practices," talks about this same dynamic and the importance of creating "ideological circles" and policing the boundaries of those circles (93–100).

For Dorothy Smith, these boundaries are not maintained simply by debate in texts; there is an organizational reality behind the documentary reality that creates an "organizational impregnability." She juxtaposes these kinds of state-controlled and mediated ideological practices as different from practices within science. She sees "scientific facticity" as "always subject to erosion, challenge, and debate, unlike state-controlled ideological practices in which there is an institutional framework in place, generating versions of reality and legitimizing those versions in an organizationally impregnable circularity. I do not think that science and state-controlled ideological practices are as different as Smith concludes, however. It seems to me that the kinds of government and foundation funding practices highlighted above certainly can have the same impact as the organizational practices that she examines in the larger social realm.[9]

Resonance with Larger Conceptual Frameworks

A second kind of legitimacy and credibility is created by producing knowledge that resonates with the dominant ideas of a society and in turn with what Gramsci (1971) called the "common sense" of the population. This common sense is a mixture of sound reasoning, experience, and folklore. It can be fragmentary, incoherent, and inconsequential as distinct from the more consistent and coherent features of the best of science. Often, legitimacy and credibility can be created in popular discourse by linking a common-sense position with the principle one is trying to convince the reader to support. In addition to appeals and resonances with common sense, I think it is important to examine how Wilson's formulation of human sociobiology, especially, resonated with the larger tenets of liberalism, and how this resonance makes it credible in popular discourse and all the more attractive to publishers.[10]

Marx referred to this larger field of dominant ideas as ideologies (Marx and Engels, 1976). Gramsci developed the concept of hegemony as a way of trying to understand the role that ideas and social relations could have on politics and social consciousness (Gramsci, 1971; see also Bocock, 1986; Femia, 1987). Foucault (1972) elaborated the concept of what he called "epistemes" to describe "the total set of relations that unite, at a given period, the discursive practices that give rise to epistemological figures, sciences, and possibly formalized systems" (191). Laclau and Mouffe (1985), borrowing from both Gramsci's notion of hegemony and Foucault's notion of discourse, take the idea further into the general field of discourse analysis and talk of discursive totalities and the "field of discursivity" (93–148). D. Smith (1990, 1999) talks about "ideological practices" and "relations of ruling" with the aim of theorizing the connections between the production of ideas and the work of institutions and social relations in a given social formation. However one conceptualizes this larger field of dominant ideas, ideologies, or discourses, and the institutional and relational framework that nurtures them, it is clear that especially Wilson and Dawkins have been able to produce a paradigm that resonates strongly with this larger field of ideas, social relations, and practices. Ruse (1999) summarizes this resonance in Wilson's work well:

> My point is that there is still something deeply cultural about evolutionary biology, even at its most mature or professional or praiseworthy level. . . .
>
> [W]hen evolutionary scientists turn to language to express their findings, the words they choose are often laden with metaphors taken from the surrounding culture....Not only do we have to contend with the fact that Ed Wilson lets his epistemic theorizing be influenced by his childhood in a militaristic society, but must also contend with the fact that Wilson thinks adaptively because we live in a society which (probably in a major part because of our Christian heritage) thinks in terms of function for organisms. And that he argues in terms of equilibrium because of American traditions which go back to the influence of Herbert Spencer....That he puts everything down to natural selection because we are today reaping the benefits of the agricultural advances brought about by the artificial selection of the early nineteenth century....And he believes in evolutionary trees because he does not live on the Canadian tundra. *Without any of these ideas, Wilson would be no more than a graduate student in search of a thesis topic. Without the metaphors of his society, his science would not exist* [author's emphasis]. (239–240)

When theory resonates with these kinds of larger and dominant social ideas, metaphors, and technical successes, then theory is more easily accessible to scientists, teachers, and the general lay public.[11] The importance of theory resonating with larger social movements is also noted by Weingart, Maasen and Segerstråle (1997). They ask and answer the question as to whether there was a connection between the general conservatism of the mid-1970s and the popularity of Wilson's *Sociobiology: The New Synthesis*.

> The most that could probably be said is that, in the conservative climate of that period, sociobiological ideas may have attracted more attention among scientists (both pro and con) and the public than had the political climate been different. The attention paid to particular topics in science by the media is a fairly reliable indicator

> of their fit with the political climate; this connection is largely constructed by the media, not just reported. (86–87)

Some of the most deterministic sociobiological arguments have had the greatest success in the popular arena. This is an arena controlled not by academics and scientists but by forces external to the academy such as publishers, editors, and marketplace demands. If we are going to understand how sociobiology has managed to thrive in both the popular and academic arenas, we need to consider the ways in which these forces have operated. These forces also are external to academic disciplines but have profound effects on the popularity, legitimacy, credibility, and, ultimately, the dominance of more deterministic versions of the relationship between biology and human behavior. Wilson, Dawkins, Barash, Dennett, and others have constructed a variant of human sociobiology that resonates strongly with and takes credibility and strength from the kindred ideas that hold sway in the larger field of discourse. They have also constructed controversial arguments that contain ideas that could engage the popular imagination. These are all things that sell books, and publishers were strongly behind the effort.[12]

Market Forces and Publishing Mandates

In *The German Ideology*, Marx and Engels (1976) first articulates the concept of "ideologies" in the manner in which I use this concept. In Marx's later work, *Capital* (1977), he moves to a more complex position on the generation of ideologies and social consciousness. In his mature work, he includes the influences that arise out of the direct involvement by ruling-class members acting in their "larger" class-for-itself interests. He also includes, however, processes whereby social consciousness emerges out of the everyday social and economic relations of capitalism. Within this framework are included actions taken by corporations and individuals to enhance their immediate economic interests, which in so doing also enhance particular sets of ideas. This appears to have happened within the publishing industry in the case of sociobiology.

Publishers constitute another set of institutional and private forces that can have a tremendous impact on the popularity, credibility, and legitimacy of a paradigm and of individual scientists. This surely has been the case in the work of Wilson and Dawkins. From the beginning, with the appearance of *Sociobiology: The New Synthesis*, we have witnessed a kind of spectacle rare in the publishing of popular works on science. According to Lewontin, Kamin, and Rose (1984):

> In the spring of 1975 a remarkable event in academic publishing took place. Harvard University Press, using the full panoply of public relations devices—including full-page advertisements in the *New York Times*, author-publisher cocktail parties, prepublication reviews and interviews on television, radio, and in popular magazines—issued a book on evolutionary theory by an expert on ants....the book *Sociobiology: The New Synthesis* and its author, E. O. Wilson, soon attained considerable celebrity. Clearly the publishers expected and promoted the book's popularity, both by their publicity campaign and by the coffee-table format of the work itself, large and lavishly illustrated with original drawings of animal societies. (233)

The authors go on to describe the many copies sold and the unusual coverage of its release in such magazines as *People*, *Reader's Digest*, and *House and Garden* They conclude: "What gave *Sociobiology: The New Synthesis* its immense interest outside of biology was the extraordinary breadth of its claims" (Lewontin et al., 1984, 234). I would agree that this surely is a part of the attraction of the book. I also would argue that popularity is, to a great extent, publisher-driven, and that publishers are looking for (and may indeed help foster) controversy,[13] accessibility, and resonance with larger ideological frameworks. These are the things that sell books. Publishers are interested in selling books, agents in getting the best deals for their writers, and writers in getting the most coverage of their work. Determinist sociobiological discourse provided, and still provides, a nice fit for all these interests.

Agents also are a key group involved in this process. A. Brown discusses the influence of agent John Brockman, who counts Dawkins and Daniel Dennett among his clients. Brockman is

> famous for his ability to command huge advances for works of popular science....These advances are so large that the publishers, to earn them back, must commit themselves to thundering publicity campaigns which have done so much to advance the Dawkinsian agenda. (A. Brown, 1999, 152)

According to A. Brown (1999), the only strong critic of sociobiology who is "on Brockman's list" is Steven Rose (152). It is likely that Rose is on Brockman's list because of Rose's 1992 book *The Making of Memory*, which is accessible and fascinating to the general public. This is a problem for critics of sociobiology, because they have tended to produce alternative conceptualizations that are much less controversial in their formulations and much less exploitable and "sexy" to the general public. Lewontin (1991) sums up this dilemma well:

> A simple and dramatic theory that explains everything makes good press, good radio, good TV, and a best-selling book. Anyone with academic authority, a halfway decent writing style, and a simple and powerful idea has easy entry to the public consciousness.
>
> On the other hand, if one's message is that things are complicated, uncertain and messy, that no simple rule or force will explain the past and predict the future of human existence, there are rather fewer ways to get that message across. Measured claims about the complexity of life and our ignorance of its determinants are not show biz. (vii)

In addition to the ways in which publishers and market influences can confer popularity, if not legitimacy and dominance, publishers also can shape the way in which science claims are presented and interpreted. When one agrees to enter into the fray and have one's ideas produced in "popular" formats, one falls prey to the direct and irreducible forces of the market and its requirements. These forces dictate the biases that must be built into the books published. Popular books about human nature must be controversial and they must ultimately be about people. Publishers of popular science works on human nature must necessarily be more committed to controversy.

A. Brown (1999) gives some sense of the inner workings of the industry in relation to sociobiology and the forces that authors must deal with when he describes the difficulties Dawkins had with German and French publishers of *The Selfish Gene*.

> The German publishers of *The Selfish Gene* put on the cover of the first edition a picture of a human puppet jerking on the end of strings descending from the word "Gen" (gene); and the French publishers a picture of bowler-hatted clockwork men with wind-up keys sticking out their backs. This is not what Dawkins thought he had meant at all. He had both covers changed, and lectured for a while using slides of the originals as illustrations of what he did not mean about genes. (30)

Publishers also must be more committed to the tastes of the buying public than to the rigor of the epistemic values of the authors. And what are these publishers committed to, exactly? A. Brown (1991) reminds us.

> I have said before that *The Selfish Gene* is a book about genes that was read as a book about people. But this is true of all the popular science of the Darwin wars. Genes don't buy books: people do; and though the book-buying public may be interested in genes, successful publishers act on the principle that the public is really trying to maximize its knowledge about people. (30)

This idea is reiterated by Weingart, Maasen, and Segerstråle (1997). Given all the dynamics I outline above, it is hardly surprising that the work of Wilson and Dawkins has been so popular.[14]

The Importance of Alternative Positions

In the battle for legitimacy, it is crucial that the strategy shift from a focus on critique to the development of alternative positions on all the fronts outlined in this chapter. Rose articulates the urgency of this task.

> The challenge to the opponents of biological determinism is that, while we may have been effective in our critique of its reductionist claims, we have failed to offer a coherent alternative framework within which to interpret living processes. (Rose, 1997, ix)

D. Smith holds that non-ideological social science must begin from people's lived experiences. In the same way, a non-ideological human sociobiology would have to begin from people's lived experiences. Also, it must begin from a set of concepts that are constructed through interactions and experiment at all levels of human lived experience. Simplistic Mendelianism is not consistent with what geneticists know about the mechanisms of inheritance and, as such, is a violation of what has been learned in the laboratory and constructed within the stringencies of the meta-values of science. S. Rose chronicles the ways in which simple Mendelianism has given way to a much more complex understanding of the ways in which genetic material interacts with cellular, systemic, cognitive, and environmental dynamics to produce organisms who act (S. Rose, 1997, 98–135). There are no individual genes for specific traits, only collections of genetic loci and complex metabolic pathways.

Disputes about the best method for analyzing human nature and human behavior are fundamentally disagreements about what is understood as the nature of the object of study—namely, human nature. Critics perceive the true objects of study to be not just the genes and genomes of individuals but individuals in all the spaces and levels of our existence—in the psychological, sociological, political, economic, and social realms that organisms create. For humans and many mammals, these creations amount to social formations that critics feel are not reducible to biological processes and that have dynamics inaccessible to those focusing on activity at the level of genes. Critics should insist on clear delineation of the many levels of reality in which human organisms operate and must also insist that the "facts" of each level be included in the reconstituting of explanations about human nature, behavior, and culture. Insistence on these conditions will help create a space for the social and will ground and constrain any evolutionary stories in actually existing genetics. Indeed, critics themselves have begun to heed Rose's advice and develop formulations that are intended to be alternatives to both the strong and weak determinist positions put forward in sociobiological discourse.

There are efforts beginning to be made that offer alternatives. These come both from within the sociobiological orbit, as I indicated in Chapter 1, and from within the ranks of its critics. These alternative conceptions of the interconnections of biology, behavior, cognition, and culture study the evolutionary, genetic, and socio-cultural roots of human and nonhuman social behavior. This work is ambitious in scope and has been able to create a small space for counter-hegemonic proposals that seek to avoid reliance on neurogenetic explanations. It is ambitious because it aims to challenge some of the fundamental suppositions of not just a human sociobiology but of all biology. These are projects that seek to redefine what organisms are, what the life sciences can know, and the ways in which this knowing can be represented. This group has a diverse membership.

Lewontin, as a population geneticist thinking about ways to frame the interactions of genes and organisms, argues against reductionism and for a special consideration for organisms and life sciences.

> It is not new principles that we need but a willingness to accept the consequences of the fact that biological systems occupy a different region of the space of physical relations than do simpler physico-chemical systems, a region in which the objects are characterized, first, by a very great internal physical and chemical heterogeneity and, second, by a dynamic exchange between processes internal to the objects and the world outside of them. That is, organisms are internally heterogeneous open systems. (Lewontin, 2000b, 113–114)

He argues further that one of the universal characteristics of organisms is the "softness of the boundary between inside and outside" (115). He also argues that the legacy of Mendel and Darwin enforces a rigid separation of inside and outside, while creating modern genetics, evolutionary and developmental biology. Yet, despite these accomplishments, Lewontin feels this legacy is now "impeding further progress" (Lewontin, 1994, 5–7).

If one takes this reworked idea of organisms seriously, then it also calls for a reworking of first principles. Lewontin's idea of "softness" calls into question many of the core concepts of genetics and evolutionary biology. Phenotype, genotype, gene, and adaptation—all would require reinterpretation.[15] There are a number of studies in recent years that argue for the idea of "phenotypic plasticity" (Coll, Bearer, and Lerner, 2004; DeWitt and Scheiner, 2004; Pigliucci, 2001). Phenotypic plasticity models seek to avoid the rigid separation of nature and nurture. This is a view of the relationship between genotype and phenotype that explores the dynamics of "environment-dependent phenotypic expression" and looks "to understand how trait values are influenced by the environment, [and] how individuals vary in this ability" (DeWitt and Scheiner, 2004, 1).[16] It is a reconceptualization of the relationship between the genetic and environmental realms to the extent that the complexity of the interactions between the two forces us to put aside the idea of using only one realm to understand human variability (Coll, Bearer, and Lerner, 2004, xviii).

Such a reworking also requires that we rethink the forms of statistical and quantitative measurement that we use in some areas of the life sciences. Rose is critical of what he calls "improper quantification" and the "belief in statistical normality" that underlie what count as facts and data when measuring things such as intelligence (S. Rose, 1997, 284–288). He makes the case in *The Making of Memory: From Molecules to Mind* (1992), that when trying to understand memory and the connection between "mind" and "brain" we can hope to "describe" the former through references to the neurobiology of the latter. We cannot hope, however, to "explain" the former simply through reliance on the latter.

Oyama (1985, 2000) has done work for the past two decades with the intent of getting past the nature/nurture dichotomy toward what she calls "developmental systems theory." This perspective employs what she calls "parity of reasoning" so that neither genetic nor environmental dynamics can be given causal priority and genetic biasing can be redressed. She also insists on "a shift from 'genes and environment' to a multiplicity of entities, influences, and environments." And, echoing Lewontin, she calls for a constructivist perspective and "a shift from central control to interactive, distributed regulation" (Oyama, 2000, 1–8). Ecologists also have initiated reconstructions of biology in the attempt to create alternative ways of representing the complexities of ecosystems and our biosphere. One of the most comprehensive and far-reaching is Vandermeer's (1996), *Reconstructing Biology*.

Other theorists have taken on the explicit task of more radical re-theorizing of organism and behavior. Webster and Goodwin (1996) present organisms as fields—"dynamically stable wholes"—when conceptualizing the processes of morphogenesis in *Form and Transformation: Generative and Relational Principles in Biology*. Mae-Wan Ho (1998) links quantum theory with organisms and behavior in a holistic participatory science in *The Rainbow and The Worm: The Physics of Organisms*. In this radical restructuring, she argues for seeing organisms as "coherent Space-time Structures." This idea of seeing organisms as coherent actors also is part of Varela's (1979) notion of organisms as autopoietic—stable, coherent, and interdependent—in *Principles of Biological Autonomy*, and in Maturana's and Varela's (1998) *Tree of Knowledge: The Biological Roots of Human Understanding*.

Feminists also have begun the project of "reinventing biology" (Birke and Hubbard, 1995). Fausto-Sterling (2000a), in *Sexing the Body*, argues for a new and complex appreciation for human organisms as the object of study and new forms of what will count as biological explanation. In so doing, she argues for a systems approach to the study of the complexities that go into the construction of human sexuality—that understands, for example, that "even labeling someone a man or a woman is a social decision" and that "our beliefs about gender affect our beliefs about what kinds of knowledge scientists produce about sex in the first place" (3).

Notes

1. For an overview of the disciplines and their involvement in the sociobiological project, see Weingart et al. (1997), *Human by Nature: Between Biology and the Social Sciences*. Within sociology the authors who have taken up the project are Lopreato (1984); Lopreato and Crippen (1999); van den Berghe (1979, 1981); and Walsh (1995). Turner, Maryanski, and Giesen (1997) discuss the reasons why Darwinian models did not take hold in sociology, citing that they have been "simply ignored as either extremist or irrelevant to sociology's interest" (29). They also cite poor early attempts at incorporating Darwinian ideas into social science and also the perception by sociologists that sociobiology is "radically reductionist" and simplistic and produces "glib, ad hoc, and easily constructed stories" (28). Recently, Fuller (2006) has proposed a "re-imagined" sociology that must come to terms with the sociobiology project. Sociobiology has fared better in anthropology. Mulder, Maryanski, and Turner (1997) claim that sociobiology and "closely related fields of human behavior and evolutionary ecology presently hold a legitimate position within anthropology" (35).
2. Shapin illustrates this process through the example of the thermometer: "Think, for example, of the physical knowledge embodied in a thermometer. To contest that knowledge would be to fight on many fronts against many institutionalized activities that depend on treating the thermometer as a 'black box.' Intercalating science or technology into larger and larger networks of action is what makes them durable" (Shapin, 1995, 312).
3. I have written extensively elsewhere on Gramsci's notion of hegemony (Schifellite, 1980).

4. I use the term "doublespeak" here to mean the act of saying one thing and meaning or intending the opposite. Some definitions hold that this is always an intentional process. I am not sure if it matters if it is intentional or unconscious if the end result is that the act is consistent and misleading.
5. See also Haraway (1976) and Paul (1998).
6. My use of the word "fixing" here is intended in the same sense as is used when one prepares or "fixes" a specimen for microscopic examination. In this process, the system being studied becomes reprocessed and "fixed," using various reagents and dyes, as a "preparation" and ultimately as an image on a slide.
7. There is a resonance here with Ruse's epistemological position presented in Chapter 4. What he calls cultural or social values can be very influential in providing the metaphors and analogies out of which many concepts emerge. The more one remains at the level of concepts, the more one is likely to forsake strong epistemic values in which knowledge is collected through experiment and intervention. The more this happens, the more one's work can be influenced by cultural rather than epistemic values. It is always a balancing act. One never escapes social influences, even as one tries to maintain the core of good science. Longino's (1990, 2000, 2002) work on epistemology also resonates well here and provides guidelines for walking this epistemological balance beam.
8. One also could examine the history of granting and funding decisions that surround the development of ideas into disciplines. One also should examine the relevant hiring and promotion decisions as well as the history of the formation of related journals, societies, and academic programs. A detailed examination of these last two areas of research is beyond the scope of this work, however.
9. One need only look at the recent case of the ignoring for over twenty years of the discovery by Marshall and Warren that *Helicobacter pylori*, in conjunction with a number of "environmental" variables, was causing gastric ulcers, and that a two-week treatment with an inexpensive and generic antibiotic would "cure" the problem. Both pharmaceutical companies whose most profitable prescriptions drugs were then being used to treat ulcers, and the many specialists whose living and professional stature came from their treatments of ulcers, were not interested in having this new treatment come to light. The doctor had to resort to creating a media event in which he infected himself with the bacteria in question and demonstrated his new ulcers, and then demonstrated the "cure" using the antibiotic. Some twenty years after his initial "discovery," the U.S. Food and Drug Administration quietly announced that indeed *Helicobacter pylori* was the chief cause of gastric ulcers. Just as quietly, pharmaceutical companies were given the right to sell what had been prescription treatments for ulcers, such as Xantac and Pepcid, as over-the-counter remedies for aiding indigestion. Finally, in 2005, Marshall and Warren received the Nobel Prize in physiology/medicine for this work. I can see little difference between an example like this and some of the examples that both Chomsky and Dorothy Smith describe.
10. In a similar fashion, Livingstone (1995) has noted that the popularity of Herrnstein and Murray's (1994) *The Bell Curve* is likely due to "the resonance of its arguments with the private world views and neoconservative political agendas of the corporate controllers of the mass media than with any scholarly merits of the book itself" (335).

11. According to the British Society for Social Responsibility in Science (BSSRS) Sociobiology Group, sociobiology's "general concepts of the world reflect a fundamental aspect of the social world in which people live and so are both easy to grasp and intuitively plausible" (BSSRS Sociobiology Group, 1984b, 132).
12. Michael Apple (1991), in "*The Culture and Commerce of the Textbook*," notes that

    "nearly seventy-five percent of the editors in college-text publishing either began their careers as sales personnel or held sales or marketing positions before being promoted to editor" and that these editors then, overwhelmingly, are driven by sales and marketing concerns (Apple, 29)
13. Segerstråle cites Mazur's (1981) contention that hints of scandal and controversy were engineered by the publisher prior to the publication of Wilson's book by planting the idea that the last chapter on human sociobiology would likely be controversial and surely bring objections (as cited in Segerstråle, 2000a, 86).
14. See, for example, the popularity of the book *Mean Genes: From Sex to Money to Food: Taming Our Primal Instincts* (Burnham and Phelan, 2000) that takes off from basic human sociobiological premises.
15. One gets a sense of this in the following passage by Lewontin: "We must replace the adaptationist view of life with a constructionist one. It is not that organisms find an environment and either adapt themselves to that environment or die. They actually *construct* their environment out of bits and pieces" (Lewontin, 1991, 86).
16. In the collection edited by DeWitt and Scheiner (2004), not only do a number of biologists discuss various aspects of this concept, but work by Sahotra (2004), a historian of biology, is also included to provide perspective on the twentieth-century historical trajectory of this idea.

· 3 ·

THE SOCIOBIOLOGY DEBATE AND THE NATURE OF SCIENCE

Legitimacy, "Reductionism," and the Nature of Science

In addition to the many levels of contested analysis engendered by the sociobiology debate, outlined in the last chapter, there is another, possibly even more profound, issue raised in the charge of "reductionism." Initially, critics charged "reductionism" when they referred to the process of reification described in Chapter 1. This is a critique of the process of reducing the social to the biological. A second meaning for this term, however, arose in the debate, and this second meaning relates to our fundamental understandings about the nature of scientific inquiry. Lewontin, Gould, and S. Rose attack the use of mathematical models and correlative statistics as representative of actual genetic entities (Segerstråle, 2000a, 175ff). In confronting what they saw and what some continue to see as sociobiology's reductionism, critics have begun to formulate ideas about the nature of science and science knowledge and the nature of the objects of study that are very different from the discourse of modernism. The overarching scientific ideals of modernism are commitments to truth, objectivity, a reductive methodology, mechanistic modeling, and quantification (Mumford, 1963, 1967).

When critics challenge sociobiology for being reductionist they are at the same time challenging an entire discourse that is grounded in a mechanistic, quantified, and reductive view of all of nature. McKinnon and Silverman (2005) argue that reductionism in sociobiological discourse and in evolutionary psychology has not been successful because it has uncovered some universal truth about human behavior but rather because these accounts "validate uniquely 'Euro-American' understandings about human nature and social life" (4). When critics challenge reductionism, they also are challenging a science and technology that has arisen in conjunction with capitalist production and that has achieved great success by reducing nature to mechanistic quantifiable bits. These core philosophical principles—that have served to legitimate the enterprise of science—have come under dispute in the past thirty years. A very different mindset has been emerging and debated along with new sets of conceptions and paradigms within biology, philosophy of science, and science studies around conceptions of epistemology, the nature of science, and the nature of the object of biology. This is a movement that seeks to abandon reductive formulations and replace them with more "complex" analyses (see McKinnon and Silverman, 2005).

Defenders have fought back strongly and the issues are by no means resolved. Defenders reacted to the charge by citing that the reducing of things to their parts is standard and fruitful operating procedure for most successful scientists. Once again, we can see that the debate has spread from the immediate topic of sociobiology to more comprehensive questions about the nature of organisms and, ultimately, to global questions about the nature of science and the nature of analysis. These questions concerning the nature of science and the limits of science knowledge are taken up in detail in this chapter.

Reductionism as Disputes About Method

Segerstråle (2000a) argues that underneath the sociobiological controversy about reductionism is a debate concerning the different conceptions about the nature of science itself (285). Reductionism has a number of distinct connotations, if not meanings, and I would agree that this is one dimension of the charge of reductionism. In this chapter I will examine this issue. The charge of reductionism in sociobiology crystallized a controversial debate over what counts as the correct method of investigation and analysis in science. The sociobiology debate did extend to questions about the nature of science; these questions led to questions of method and ultimately to what will count as good science. Wilson and others situated their project firmly within what they present as the methods of "normal" science, which conjures a collateral legitimacy for their work. They also claimed reductionism only as a method and, at the very least, pay lip service to notions of building complex arguments out of this reductive method. Wilson has defended himself against the charges of reductionism by defining reductionism as a method used by all scientists to mean the teasing apart of the object of study into constituent elements for the purpose of analysis and understanding. In this, he understood himself to be doing normal, "good," and uncontroversial science. For Wilson (1978),

> the heart of the scientific method is the reduction of perceived phenomena to fundamental, testable principles. The elegance, we can fairly say the beauty, of any particular scientific generalization is measured by its simplicity relative to the number of phenomena it can explain. (11)

In this manner, Wilson is not alone in following a method based on experiment, observation, dissection, and isolation of variables that can be manipulated and from which conclusions, facts, and theories can be drawn. In this manner, he is able to claim natural science respectability and implicitly to position the work in opposition to "mushy" and "soft" forms of analysis carried out in the social sciences.

Furthermore, scientists such as Wilson and Dawkins argue that they make a distinction between a commitment to a "reductionist method" and a commitment to "reductionist metaphysics" (Segerstråle, 2000a, 285). In *On Human Nature*, Wilson (1982) is clear:

> [R]aw reduction is only half of the scientific process. The remainder consists of the reconstruction of complexity by an expanding synthesis under the control of laws newly demonstrated by analysis. This reconstitution reveals the existence of novel, emergent phenomena. When the observer shifted his attention from one level of organization to the next, as from physics to chemistry or from chemistry to biology, he expects to find obedience to all the laws of the levels below. But to reconstitute the upper levels of organization requires specifying the arrangement of the lower units and this in turn generates richness and the basis of new and unexpected principles. (11–12)

Critics did not see this as a successful defense. What critics such as Lewontin, Gould, and Rose wanted was for Wilson, in effect, to be true to his commitment to move from reduction to complexity. This is what is seen as missing from Dawkins' and others' work. I do not believe that critics such as Gould and

Lewontin argue against experiment and the deconstructing of phenomena into their constituent parts for analysis. The essence of the charge of reductionism is about Wilson not being true to the principles that he lays out for himself in the passage quoted above. What Lewontin, Gould, and others object to is that Wilson does not accord each level of human reality and its subsequent mode of analysis "equal time" or indeed any space in his human sociobiology.

Segerstråle has argued that Lewontin and Gould present a contradictory critique of Wilson's reductionism (2000a, 288–290). She claims that they first argue that Wilson tries to reduce complex human behaviors to numbers and atomized genetic components, while at the same time they insist that Wilson stick to the simple experimental facts of molecular biology (275–294). Critics such as S. Rose do not oppose reductive methodology per se. They do, however, oppose it when reduction as a method becomes a metaphysical position, and they do oppose what metaphysics says about the study of living organisms and the organisms themselves.

What critics object to is what Rose calls "theory reduction" and "philosophical reduction" (S. Rose, 1997, 80–97). By theory reduction he means the drive in science "to try to embrace a maximal description of the world within the minimum possible number of laws and variables" (80). Levins and Lewontin (1985) argue that "the dominant mode of analysis of the physical and biological world...has been Cartesian reductionism (269). They characterize this Cartesian mode as arising from Descartes' *Discours*.

In this Cartesian paradigm, the world is viewed "as a clock," in which "phenomena are the consequences of the coming together of individual atomistic bits, each with its own intrinsic properties, determining the behavior of the system as a whole" (1–2). According to Levins and Lewontin, this paradigm has four ontological commitments that shape the process of knowledge creation:

1. There is a natural set of units or parts of which any whole system is made.
2. These units are homogeneous within themselves, at least insofar as they affect the whole of which they are the parts.
3. The parts are ontologically prior to the whole; that is, the parts exist in isolation and come together to make wholes. The parts have intrinsic properties, which they possess in isolation and which they lend to the whole. In the simplest cases the whole is nothing but the sum of the parts; more complex cases allow for interactions of the parts to produce added properties of the whole.

> 4. Causes are separate from effects, causes being the properties of subjects, and effects the properties of objects. While causes may respond to information coming from the effects (so-called "feedback loops"), there is no ambiguity about which is causing subject and which is caused object. (This distinction persists in statistics as independent and dependent variables.) (1985, 269)

Ultimately, for Levins and Lewontin, the error of this Cartesian reductionism "is that it supposes that the higher-dimensional object is somehow 'composed' of its lower-dimensional projections, which have ontological primacy and which exist in isolation, the 'natural' parts of which the whole is composed" (1985, 271). Critics such as Lewontin, Rose, and Kamin (1984) and S. Rose (1980, 1995, 1997) object to this process—especially when it is used to argue for a strong genetic determinism in humans.

Levins and Lewontin (1985) have developed a comprehensive alternative to Cartesian principles in *The Dialectical Biologist*. These authors propose a holistic perspective, which can inform a new philosophy and methodology in biology. This holism is informed by what they call a "dialectical view" of reality (272). This dialectical view carries a number of foundational principles.

> The first principle of a dialectical view, then, is that a whole is a relation of heterogeneous parts that have no prior independent existence *as parts*. The second principle, which flows from the first, is that, in general, the properties of parts have no prior alienated existence but are acquired by being parts of a particular whole....
>
> A third dialectical principle, then, is that the interpenetration of parts and wholes is a consequence of the interchangeability of subject and object, of cause and effect....
>
> Because elements recreate each other by interacting and are recreated by the wholes of which they are parts, change is a characteristic of all systems and all aspects of all systems. That is the fourth dialectical principle....(273–279)

Through the use of this dialectical view, processes, which upon first inspection might be thought of as either deterministic or random, in a mutually exclusive sense, become interoperable. For Levins and Lewontin (1985), "completely deterministic processes can generate apparently random processes"; and "random processes may have deterministic results" (284). Ultimately, random and deterministic processes interact to give "results in evolution that are different from the consequences of either type of process acting alone" (284). According to Levins and Lewontin, this dialectical view has "implications for research strategy and educational policy as well as methodological prescriptions" (286).

These include commitments to historical analysis and to the principles of universal interconnectedness, heterogeneity, the interpenetration of opposites, and the integration of levels of analysis (286–288).[1]

S. Rose also does not think this Cartesian enterprise is a useful goal in biology and feels that it creates tendencies toward producing what he sees as the much more dangerous philosophical or metaphysical form. In this conception, nature is arranged in a hierarchical pyramid with physics and its object of study at the bottom or most fundamental level and sociology and its objects of study at the top and most epiphenomenal level. According to S. Rose (1997),

> because science is unitary, and because physics is the most fundamental of the sciences, then an ultimate [theory of everything] will be able to reduce chemical theory to a special case of physics, biochemistry to chemistry, physiology to biochemistry, psychology to physiology, and ultimately sociology to psychology—and hence to physics (83)

In addition to critiquing reductionism, and to beginning the process of creating philosophical and methodological models that move beyond Cartesian paradigms in their understanding of the ways in which parts and wholes interact, questions about the nature of science also extend to questions about what will count as good science.

Good, Bad or Socially Influenced Science?

One of the interesting disagreements within the human sociobiology debate has been around the issues of evidence and incaution. Both sides are examining the same world. Yet sociobiologists feel that they are justified in their speculations, while critics feel that there is no evidence to support these speculations. Essentially, critics claim a completely unsubstantiated leap from discussions of insect and other nonhuman societies directly to human societies. For Kitcher (1985), "the dispute about human sociobiology is a dispute about evidence" (8), and "while claims about nonhuman social behavior may be carefully and rigorously defended, the sociobiologists appear to descend to wild speculation precisely where they should be most cautious" (9). Gould (1980b) echoes these criticisms and reminds readers that sociobiology's claims are unsupported and its methodology flawed (285).

When critics saw what appeared to be "wild speculation" leading to claims that many of the social inequities in society are genetically determined and, therefore, not easily changed, they assumed that these proponents of sociobiology had a conservative political agenda. This bothered many who began to worry that the "authority of current scientists" would encourage the acceptance of "politically harmful falsehoods" based on "inadequate and misleading evidence" (Kitcher, 1985, 6). And so some large part of the early critiques were based on the potentially damaging impact that sociobiology could have on the policies aimed at reducing social inequities.

Critics charged that especially when dealing with discussions of human nature that may be picked up and used to influence public policy, one must be especially careful and cautious in the pronouncements made and in the kind of speculations engaged in. Kitcher (1985) contended that if the human sociobiology project were highly speculative and the genetic determination so qualified (as even Wilson (1982) sometimes seems to indicate in *On Human Nature*), then why make premature statements about the implications for social policy? As Kitcher (1985) has pointed out:

> Those who form social policies often look to the findings of the sciences for guidance...when scientific claims bear on matters of social policy, the standards of evidence and of self-criticism must be extremely high. (3)

Ultimately, Kitcher points out that "the sociobiologists appear to descend to wild speculation precisely where they should be most cautious" (9). Wilson, Lumsden, Barash, and other sociobiologists simply replied that truth and its quest must not be filtered regardless of whether we like what we find (Wilson, 1975; see also Alexander, 1979; Lumsden and Wilson, 1983). Gould counters, however, that he has no desire to suppress truth. Rather,

> in stating that there is politics in sociobiology, I do not criticize the scientists involved in it by claiming that an unconscious politics has intruded into a supposedly objective enterprise. For they are behaving like all good scientists—as human beings in a cultural context. I only ask for a more explicit recognition of the context and, specifically, for more attention to the evident impact of speculative sociobiological stories. For example, when the *New York Times* runs a week-long front page series on women and their rising achievements and expectations, spends the first four days documenting progress toward social equality, devotes the last day to potential limits upon this progress, and advances sociobiological stories as the only argument for potential limits—then we know that these are stories with consequences. (Gould, 1980a, 263)

Wilson's equivocations coupled with the kinds of doublespeak discussed in Chapter 2 have led many critics to charge that human sociobiologists are social and political conservatives and that they have deliberately or naively constructed human sociobiology so as to support and rationalize contemporary social inequalities. Critics accused Wilson of having his science influenced by neoconservative political beliefs. According to Lewontin, Rose, and Kamin (1984), Wilson has identified himself in a media interview with "American Neoconservative libertarianism, which holds that society is best served by each individual acting in a self-serving manner, limited only in the case of extreme harm to others" (264).

Lewontin et al., (1984) also contend, in support of their charges of a sociobiological political agenda that

> the development of the literature on sociobiology since 1975...including Wilson's own *On Human Nature*, leaves little doubt that the problem of human nature is at the center of sociobiological concerns. (243)
>
> Sociobiology and the pop ethologies are forms of human nature theory that in some aspects characterize all political philosophy. (240) In hypostatizing entrepreneurial bourgeois society, sociobiology is a direct intellectual descendent of Thomas Hobbes's *Leviathan* of 1651. . . .
>
> The influence of Hobbes's thought on sociobiology comes not directly, but through the intermediary of Darwinism and social Darwinism. (241)

[and, ultimately]

> There is really nothing that separates the program or specific claims of the social Darwinism of the 1870s from the Darwinian sociobiology of the 1970s. (243)

These authors were not the only ones to notice what appears to be the political intent of human sociobiology. As Kitcher (1985) put it, in summarizing van den Berghe's (1979) *Human Family Systems*: sociobiology's truths about human society seemed to boil down to asserting "that the family is here to stay, that women are better with the kiddies, that men are turned off by aggressive females and that politics is a man's game" (5–6). Kitcher also contends that individual sociobiologists may not have racist, sexist or classist agendas. Yet "because they stress the genetic basis of behaviors, many sociobiologists seem to be endorsing a strategy of linking behavioral differences to genetic differences, and this strategy encourages the denigration of particular racial and social groups" (1985, 6). And as I indicated in Chapter 1, the charges of undeclared political agendas and lack of evidence for the claims made have not abated. McKinnon (2005) has leveled these same kinds of charges at evolutionary psychology—a new branch of psychology seen as one of the progeny of Wilsonian sociobiological discourse.[2]

Wilson and others have consistently countered these charges in two ways. In reply to the charges of conservative political influence, they deny that this is so and show that human sociobiology has appealed to the left as well as to the political right (Fuller, 2006; Singer, 2000). Wilson argued that left-wing intellectuals like Noam Chomsky and Herbert Marcuse were interested in sociobiology (Wilson, 1978, 301–302). Segerstråle (2000a) attempts to add to this argument by noting that a left-wing scientist such as Salvatore Luria chose not to see sociobiology as necessarily political or worth his time in arguing over its principles. In support, Weingart and Segerstråle (1997) point out Masters' (1982) observation that, "sociobiological thinking and its central idea of inclusive fitness theory is clearly part of the individualist tradition in Western thought, but this tradition may be used for either conservative or progressive purposes." (85). Weingart and Segerstråle also cite van den Berghe (1981) in disputing the political roots of sociobiology.

> Actually, a review of the politics of leading sociobiologists would lend more credence to the contention that sociobiology is a Communist conspiracy: J. B. S. Haldane, who is generally credited for having first hit on the notion of kin selection, a theoretical cornerstone of sociobiology, was a leading member of the British Communist Party; so was John Maynard Smith. E. O. Wilson and most other leading sociobiologists are left-of-centre liberals. "Racist" Trivers is even married to a Jamaican and is heavily involved in radical black politics. (Weingart and Segerstråle, 1997, 86)

One must be careful here, however, with keeping basic ideas in sociobiology separate from the human sociobiology enterprise. Critics seemed careful to direct their comments at the project of human sociobiology, not necessarily at specific principles common to nonhuman sociobiology such as inclusive fitness. Likewise, Ruse (1979) contends that Maynard Smith is someone who "flatly dissociates himself from human sociobiology and who denies that his ideas have any relevance to the human case" (53).

The second defense used by Wilson against the charges of conservative politics influencing the development and claims of human sociobiology has been to write off critics as Marxists and leftists with their own political agenda. He identifies them as Marxists, and therefore "propagandists" with biases that taint their ability to do objective science, while claiming for himself the nonbiased middle ground as I discussed in Chapter 2. In the 25th-anniversary edition of *Sociobiology: The New Synthesis*, published in 2000, Wilson pushes this point.

> Who were the critics, and why were they so offended? Their rank included the last of the Marxist intellectuals, most prominently represented by Stephen Jay Gould and Richard C. Lewontin. They disliked the idea, to put it mildly, that human nature could have any genetic basis at all.... Theirs was the standard political position taken by Marxists from the late 1920s forward: the ideal political economy is socialism, and the tabula rasa mind of people can be fitted to it.
>
> In the 1970s, when the human sociobiology controversy still waxed hot, however, the Old Marxists were joined and greatly strengthened by members of the New Left in a second objection, this time centered on social justice. If genes prescribe human nature, they said, then it follows that ineradicable differences in personality and ability also might exist. Such a possibility cannot be tolerated. At least, its discussion cannot be tolerated, said the critics. (vi)

Wilson then goes on in this foreword to claim victory:

> The argument for a political test of scientific knowledge lost its strength with the collapse of world socialism and the end of the cold war. To my knowledge it has not been heard from since. . . .
>
> Capitalism may yet fall—who can predict history?—but, given the overwhelming evidence at hand, the hereditary framework of human nature seems permanently established. (vi)[3]

Segerstråle (2000a) has written quite extensively in *Defenders of the Truth* in defense of a more sympathetic reading of Wilson's overall work and position. While Segerstråle holds that "Wilson had been one of the pioneers in storming the culturalist bulwark" (317), she does not see his intentions as politically motivated in the way that critics charged. Likewise, she sees his and Lumsden's later attempts at coevolutionary theory as sincere efforts to give cultural influences a place within the paradigm.

In addition to the charges that Wilson and others were influenced by their personal political beliefs, critics also charged that sociobiology as a paradigm is influenced by larger dominant discourses or ideologies.[4] They point out that science is a social construct and, as such, is subject to the same influences as any other social activity. As I indicated in Chapter 2, critics have charged that sociobiological theory derives its credibility and legitimacy from its resonance with dominant ideas and discourses prevalent in the larger society. Likewise, critics also charge that sociobiology has been "tainted" by the implicit and explicit adoption of ideas and metaphors imported from the larger social discourses, technologies, and social relations.

Lewontin argues that the modern tendency toward a Victorian social Darwinism is driven by ideas and metaphors imported from Victorian ideologies

(Lewontin, 1991). Haraway argues that biology in general and sociobiology in particular represent a further elaboration of techno-rational values and metaphors generated in twentieth-century industrial capitalism, liberalism, and industrial society (Haraway, 1991, 43–68). Young's (1985) analysis of the social influences on Darwin's own formulations of the idea of selection sits as a groundbreaking work in the history of evolutionary theory. There also has been excellent work done on the connections between dominant political ideologies, such as liberalism, and the fields of evolutionary psychology and genetics in the twentieth century.[5] Finally, according to S. Rose (1997), there is a coherent critique that is well grounded philosophically and analytically. This critique "sees modern science as the inheritor of 19th-century mechanical materialism, itself tightly linked ideologically to a particular phase of the development of industrial capitalism" (73).

Critics argue that these larger social influences affect the accuracy and validity of Wilsonian human sociobiology and the general determinist project. While this may be an accurate assessment, it has taken critics down a very difficult path. Engagement with questions about the legitimacy and success of science knowledge and science paradigms almost immediately invokes questions about how one assesses the "truth" or at least the accuracy of claims being made. According to a more neopositivist epistemology, social influences are factored out of ultimately accepted science claims by the systems of evaluation and review in place within disciplines (Richards, 1987, 559–593). And even if social influences stay, they likely influence the choice of research area rather than the "truth" content of any given research chosen. In this scenario, there is either "good" or "bad" science. If scientists have followed the strong epistemic rules, the science is ultimately "good" and "truthful."

Indeed, this is exactly the position that Wilson and others take against critics. Wilson clearly sides with neopositivists in his formulation of sociobiology as a "hard science" committed to deciphering core principles using established methodological principles (Segerstråle, 2000a, 2000b). Early promoters of sociobiological discourse simply say that they have "gotten it right," that they discovered the truths linking genetics, evolution, and behavior and have been rewarded for this. These defenders argue that even though there may be social influences, this does not necessarily mean that the scientific positions arising through these influences are not accurate. For supporters of the sociobiology enterprise, it usually does not matter where the inspiration or metaphors for the work are derived as long as one holds that the precepts and practices of science ensure that truth can be winnowed from ideologies. These supporters still hold out the possibility of creating "truth."

A straightforward neopositivist critique of human sociobiology would cite Wilson and others for blatant contraventions of the tenets of "good science" and for other kinds of methodological errors, while maintaining the possibility that good materialist accounts of the phenomena in question are possible. Critics invoking a social influence or social constructivist model, however, cannot rely on neopositivist theory and therefore its version of what can count as "good science." Critics have hoped to undermine the core ideas in sociobiology by charging social and ideological influences. At the same time, they also began to confront the need to clearly delineate what could count as "good" science in opposition to Wilson's "bad" science and, in general, also to delineate what the epistemological basis should be for making scientific claims.

This has led critics onto unstable epistemological ground. If all knowledge is socially influenced, then what about the critics' positions? Defenders of determinism have been able to deflect critics either by claiming support for a neopositivist position or by challenging the "veracity" of critics' positions because of the socially constructed nature of the critics' own work. Constructionism implies giving up a privileged epistemological position and thereby undermining critics' rights to make claims as well. Are critics not as susceptible to social and political influences as the sociobiologists they critique? Conversely, insistence on at least a privileged method in science limits the critics' charges ultimately to those of "bad" science or improper method, but then one loses the analysis of social influences on the formulation of human sociobiology. As Richards put it, "[W]hat criterion shall we employ to choose between the ideology of these sociological historians and that of the sociobiologists they condemn?" (Richards, 1987, 557) Longino quotes Haraway and summarizes the problem in this manner:

> As Donna Haraway has observed in a review of several collections of essays on sociobiology and hereditarianism, to simultaneously adopt an analysis of observation in science as theory—or paradigm- determined while asserting the incontrovertible existence of any fact is to embrace paradox. Underlying her critique is the idea that if observation is theory determined, then we can have no confidence that what appears to be fact in the context of one theory will remain so in the next. Indeed if sexist and racist science is bad science that ignores the facts or fails to treat them properly, this implies that there is a good or better methodology in the world that will steer us away from biased conclusions. On the other hand if sexist science is science as usual, then the best methodology in the world will not prevent us from attaining those conclusions unless we change paradigms. (Longino, 1990, 11)

At the same time as critics such as Lewontin, Rose, Gould, and others began to think about these issues, those working in the fields of science studies were beginning to deal with the same issues. The debate within sociobiology about the legitimacy of its claims is echoed in a larger debate that has been ongoing and that concerns the general nature of science and its epistemological underpinnings. Critics' focus on the social construction of sociobiological knowledge has invoked this much larger debate about the nature of all science knowledge. This debate calls the nature of science into question. In the past four decades, this general debate has occupied those doing studies of science in philosophy, history, and sociology. This larger debate is very much a work in process. It has not been resolved but has instead created a continuum with very polarized constructionist and neopositivist positions at the two ends.

What ultimately is at stake on this continuum of positions is the place that science has as a generator of knowledge. At the neopositivist end of the continuum one would ask if even when factoring in social influences, does science still produce a unique knowledge that is a "true" representation of reality? Or, at the other end of the continuum, is the kind of knowledge that science produces equal only to other kinds of knowledge, be they religious, speculative, or common-sensical? A kind of Kuhnian incommensurability has opened up in this debate, with people at both ends unable to speak to one another about the gulf between their positions. The conflict came to be known as "The Science Wars" (Hull, 1988; Ross, 1996; Ruse, 1999; Segerstråle, 2000b).

Critics of human sociobiology reflected this unsettled and contradictory debate in their own work. Both critics of sociobiology and those doing work in science studies have had difficulty developing workable answers to these questions. What has sometimes happened in the sociobiology debate is that the same critics employ, at different times, both positivist and constructionist positions. This leads to the appearance that critics sometimes appear to be arguing both for sociobiology as "bad science," that is, in violation of neopositivist methodological norms, and simultaneously arguing that social factors influence sociobiology's formulations in ways that preclude the possibility of "truth."[6]

In the previous two chapters, I looked at the ways in which the debate about sociobiology became debates within the larger conceptual frameworks of evolutionary theory and genetics. In similar fashion, debates about sociobiology's "reductionism" ultimately became debates about the nature of science and the status of science knowledge. In the last chapter, I highlighted the need for critics to move beyond a critique of reductionism toward the formulation of alternative conceptual positions. Likewise, in order to find a way out of this epistemologically dense forest, critics must develop alternative epistemological positions that theorize how we understand the influence of social forces on science claims and revise what will count as good science in this context. On the other hand, those seeking to promote the legitimacy and popularity of the more Wilsonian human sociobiology must work to maintain the boundaries of debate so as not to allow these alternative conceptions into the debate. The rest of this chapter includes a discussion of the history of the development of what has come to be called the "social construction of science" movement and the ensuing science wars and of what may be considered a reasonable epistemological "middle ground." This "middle ground" can be useful in evaluating Wilsonian sociobiological claims and in the analysis of the biology textbooks that is taken up in Part II of this book.

The "Science Wars" and the Nature of Science Knowledge

The twentieth century has seen the proliferation and ascendancy of various forms of social constructionist arguments in the history, philosophy, and sociology of science. Philosophers, historians, and sociologists of science since the 1960s have done increasing numbers of studies that show that science is not the isolated, objective, empiricist enterprise that it had been considered (Shapin, 1995; Zuckerman, 1988). Instead, these researchers began to develop a picture of science as a human activity that is influenced by individual, social, political, and economic contexts—as are all other kinds of human activity. This development in the social studies of science has spurred great controversy that has culminated in, as I mentioned above, what has become known as "The Science Wars." The issue, at this point in the debate, is not that the sociopolitical context influences how science is done. The issue is about the impact of this social context on the product of science—its facts and theories. Given these social influences, the issue also becomes one of developing an evaluative constructionist method for evaluating controversial claims in science. In the next section, I trace some of the important historical developments in the project of understanding this impact. This is not meant to be an exhaustive account, and I do not set myself up as an expert on this affair; instead I intend only to provide the reader with a grounding in the important issues.

The Historical Context

While philosophers have long debated the status of what we can know about reality and about how we come to know the world, it is John Stuart Mill, influenced by Auguste Comte, who clearly articulates what we would now call the hypothetico-deductive method (Oldroyd, 1986,154) and positivist science. This method was designed to help ensure that science develop a series of practices that when followed are able to produce viable, consistent and "true" knowledge about the world. However, from the early twentieth century onward, there has been a steady erosion in the philosophical belief that science can produce objective knowledge.

In part, philosophers of science became focused on issues of truth and objectivity because of developments in relativity theory and quantum mechanics in the early part of the century. Relativity theory challenged the idea of a privileged point of observation and showed that at relativistic speeds, the point of observation was extremely important. Quantum mechanics added an element of fundamental uncertainty to science knowledge claims. Quantum mechanics began to challenge the idea that reality had a kind of fixity that had been assumed, and it also claimed that the act of observation was influential in determining how reality appeared. This work was complemented by developments in psychology that were beginning to view perception not simply as a passive process of information input but as an active process of selecting and shaping stimuli.

Gradually science was coming to be seen as a complex process in which our assumptions about our science-based interactions with the world were called into question. Scientists and philosophers were beginning to come to the conclusion that reality may not be as fixed as had been assumed, and the entire process of discovery and articulation may be subject to many influences outside the traditional domain of science. Albert Einstein came to see the roles of conceptual invention and active observation in science. According to Gregory (1988), for Einstein,

> [it] is the theory that decides what we can observe (93)....Science is the attempt to make the chaotic diversity of our sense-experience correspond to a logically uniform system of thought (43)....Knowledge cannot spring from experience alone, but only from the comparison of the inventions of the mind with observed fact. (5)

Ultimately, Einstein expressed a form of neo-empiricism and correspondence or copy theory of truth (Oldroyd, 1986, 273) in which human invention, conception, and action played a role in producing the models and metaphors with which to think about reality.

His contemporaries in quantum mechanics were more radical. The subatomic object of quantum mechanics seems to be able to behave in wavelike or particle-like ways, depending upon its environment—upon "the experimental situation, that is, on the apparatus it is forced to interact with" (Capra, 1983, 79). Niels Bohr introduced the idea of complementarity, in which particles and waves were two complementary descriptions of the same thing. Quantum mechanics' implicit radical view of reality is one in which there are tendencies of existence (80) that can be expressed as probabilities (for example, Heisenberg's Uncertainty Principle). Quantum theory came to see reality as closely interconnected and having strong extra-local relationships. In the epistemological positions of early relativity theorists and quantum theorists, one can see most of the facets of the debate that would later rage in the philosophy of science.

Some responded by attempting to preserve and reconstruct a viable neo-empiricist position. The most important of them is Karl Popper (1963), whose criterion of falsifiability for a time appeared to give back to scientists a large degree of control in deciding what could count as scientific knowledge. Others moved toward and sometimes down the path of epistemological relativism. However, few pre-World War II philosophers and scientists could bring themselves to a relativist position that removed the special status of science and its method. Early twentieth century sociologists of science were quite willing to accept the social dimension and influence of most forms of knowledge, but they too tended to reserve special status for at least the "hard" sciences. Developments in the sociology of knowledge and the sociology of science were similar to those in philosophy and history and were fraught with the same tensions and contradictions. Marx and Engels (1976), and later Marx (1977) (whose work pioneered what has become the sociology of knowledge), seemed to stop short of allowing the social context to shape the form and context of science knowledge. At best, one can say Marx was ambiguous on this point. Emile Durkheim (1915) did maintain that social relations influenced the form of knowledge generated, but he exempted science from his considerations (Mulkay, 1979, 3). Mannheim was quite radical in his insistence that social relations influenced the form and context of knowledge. Yet he stopped short of this radical position for mathematics and natural science (Mannheim, 1936, 35) and ultimately maintained the distinction between natural science and other forms of knowledge (Mulkay, 1979, 11). Merton (1970, 1973) as well reserved for science and its claims a special status and ultimate independence from social context and influences.

Those wishing to understand the ways in which the larger social setting and ideological framework influenced science knowledge began to study this problem in the early decades of the twentieth century. This work was being stimulated from many directions. One was the emergence of anticapitalist states and ideologies in which there was a desire to understand the place of science as a social phenomenon in the development and reproduction of capitalism (Rose and Rose, 1982). Researchers such as Hessen (1931), Fleck (1979), and Bernal (1939, 1971) pioneered these more radical investigations into the social relations of science. Hessen (1931), in his work on Newton's theories, did attempt to show how capitalist development and bourgeoisie interest decisively shaped Newton's formulation. But even he, in the end, fell back on the Marxist-Leninist notion that "genuine scientific knowledge of the laws of historical process" can be produced (211). Fleck (1979), whose work received little interest when first published in 1935, presaged the work of Kuhn (1962) and more recent studies by Latour and Woolgar (1979). Fleck's project, as its title implied, was to study *The Genesis and Development of a Scientific Fact.* He employed the concepts of "thought collective" and "thought style" that were similar to Kuhn's notions of paradigm and scientific community. He attempted to demonstrate that "[any] fact is possible as long as—indeed only if—it fits the accepted thought style" (157). According to Trenn, for Fleck, "the thought style is a social product; it is formed within a collective as the result of social forms. The circumstance links problems of natural science with those of sociology and especially the sociology of thought" (Trenn, 1979, xviii). Fleck saw knowledge as undergoing change as thought styles changed. And, as Trenn and Merton (1979) point out, what was revolutionary about Fleck's theory of knowledge was its application to scientific cognition (163). Yet even Fleck leaves his epistemological position vague. He seems to deny the "absolute and unconditional character of science" but still believes that science produces cumulative knowledge (163). He stops short of extreme relativism.

All in all, as Mulkay has indicated, despite important differences in various works, pre-war sociology of science held,

> first, that science flourishes in large scale industrial (capitalist) societies and that within such societies, scientists create distinct communities which regulate production of certified knowledge; second, that although the rate of growth, the focus of attention and the use made of scientific knowledge is in large measure socially determined, its context is independent of its social influences; and third, that scientific research

> communities are likely to have special social characteristics which reduce the impact on members' technical work of such distorting factors as bias, prejudice, and irrationality, and which are, therefore, crucial in enabling scientists to generate objective knowledge. (Mulkay, 1979, 10)

Mulkay (1979) has suggested that even Mannheim, who developed a radical epistemological position for most knowledge, stopped short of including natural science. It is interesting to note that physicist Max Planck was more radical than these philosophers, historians and sociologists of science in his estimation of the influences of social forces on the acceptance of new scientific theories. This is evident in his statement: "A new scientific truth does not triumph by convincing its opponents and making them see light, but rather because its opponents die, and a new generation [grows] up that is familiar with it" (Gregory, 1988, 71).

Rose and Rose (1982) point out that it was not until after World War II that studies of the external or social dimension of science really developed. They attribute this to the fact that the forced intervention into science by states during World War II made the connection more obvious (16–17). These studies began to look at the relationship between existing facts, experiments, and theories and the acceptance of new facts, experiments, and theories.[7] These developments in the post-war study of science pushed epistemology and sociology of science in the direction of a stronger social constructionist position and a stronger relativism. This also has meant accepting the notion that facts are constructed and that observation and theoretical frameworks influence their production.

Hanson's (1958) work in philosophy was crucial in this area. Hanson was greatly influenced by Ludwig Wittgenstein and by experiments in Gestalt psychology that demonstrated the importance of context and perspective in sense perception. From these he began to develop a more radical and relativist philosophy of knowledge and perception, the basis of which tends to be accepted by most theorists today. Most would agree that, as Hanson said, no one simply "sees" an event; they "see as" or "see that" and construct the perception in relation to their position and context. The variables involved are both individual and social. This shift in our understanding of the dynamics of perception allowed for more questioning of the role that existing theory and larger ideological frameworks have in determining new research and the development of new facts.

The watershed piece that focused and represented the growing interest in social and contextual influence on science theory is Kuhn's (1962) seminal work, *The Structure of Scientific Revolutions*. In that work, Kuhn set himself against Popper and others attempting to save traditional empiricism. Kuhn did not see the progress of science knowledge as a simple case of cumulative development in which a proper method mediates the results. He was much influenced by Hanson's work and instead saw the theory-ladenness of facts, the importance of perception and interpretation, and the inability of those holding different paradigms to see the same world and to communicate their observations to one another. In this work, he seemed to employ a radical epistemology (although implicitly) and also gave the social dimension a large role in fostering the key motor of paradigm change—scientific revolutions. Even though Kuhn would later appear to retreat from his initially more radical epistemological position, he still argued that different "paradigms" look at different worlds. *The Structure of Scientific Revolutions*, crystallized the terms of the debate, set the stage for a new generation of philosophers and historians of science, and gave impetus to studies in the sociology of knowledge and sociology of science.

At the same time, others were also adding social dimensions to science in their work. Polanyi (1966) developed the concept of "tacit knowledge," i.e., that almost invisible knowledge that scientists hold and share and that shapes their research and their conclusions. Tacit knowledge is learned by scientists as part of their training. It is in this training that they in fact learn the boundaries of their field and its "interesting issues." Although Polanyi would not have been happy with strong relativist interpretations of science knowledge, tacit knowledge added an important social dimension. As well, Lakatos (1968), a student of Popper's, examined the process of conducting science, and in so doing examined the unstated rules of experiment that guide research. In so doing, Popper and Lakatos both attempted to articulate ways in which competing theories can be evaluated.

The general trend in the philosophy of science in the latter part of the twentieth century, as Oldroyd (1986) points out, was that "the positivists' picture of science [had] indeed come to be seriously questioned" (292). On balance, this has left the philosophy of science somewhat polarized. The Popperian wing has sought not to recreate positivism but to save science from relativism. Popper developed many methodological devices (e.g., falsifiability) in attempts to shore up the special status of science. None of these, however, appears to have endured (297–317). Popper and more recent philosophers also have sought to develop an evolutionary notion of science theory, wherein a kind of natural selection process decides or selects for theories that are the fittest (Richards, 1987). Yet such a theory begs as many questions as it purports to solve. Such attempts have not been able to refute the challenges of social constructionist and radical epistemological critiques.

One of the most controversial and radical of these constructionists was Paul Feyerabend, who began publishing in the 1970s. He took a very strong and contested stance stating that the scientific method had no privileged position and, in fact, there was no one scientific method. Not only was there no one method in science but some of the "progress" of science, he said, came out of irrational thinking. His appears to be a kind of skeptical empiricism whose relativism "is not about concepts...but about human relations" (Feyerabend, 1987, 83). According to Feyerabend,

> critics of the idea that scientific debates are settled in an objective manner do not deny that there exist "means of deciding" between different theories. On the contrary, they point out that there are many such means; that they suggest different choices; that the resulting conflict is frequently resolved by powerplay supported by popular preferences, not by argument; and that argument at any rate is accepted only if it is not just valid, but also plausible, i.e., in agreement with non-argumentative assumptions and preferences....(80–81)
>
> ...The events and results that constitute the sciences have no common structure: there are no elements that occur in every scientific investigation but are missing elsewhere....Successful research does not obey general standards....The most we can do [for scientists facing some concrete problem] is to enumerate rules of thumb, give historical examples, present case studies containing diverging procedures, demonstrate the inherent complexity of research, and so prepare them for the morass they are about to enter. (281)

Ultimately, Feyerabend (1987) argued for a kind of Millsian pluralism, without Mills' positivist conviction in which it is important that as many versions of reality as possible are allowed to flourish in the hope that this will provide checks and balances over any one statement of "truth" (34).

Feyerabend's work, however, contains epistemological ambiguities. At one level he seems to be saying that choosing between competing theories is always a social process. We can never rise above this to a privileged position; we can only compare the competing positions openly and play them off one another. Yet at other points in his argument he refers to great scientists and to successful research but gives the reader no clue as to how this status is conferred; and this leaves the reader without some sense of what good science is, beyond the statement that it is similar to great art. Is great science, then, simply what is capable of capturing the popular and scientific imagination? Is it that which seems to lead to technological progress? Is it that which can predict events with regularity? Are all of these elements crucial, or are none of them crucial?

At the same time as Feyerabend was writing, excellent socio-historical studies were also being developed. One very important author whose work has influenced much of the work that has followed is Foucault. He creates a strong interpretation of social influences in which he comes to see knowledge/power as an inseparable social duality. Foucault's (1970, 1972) development of the techniques of archaeology and genealogy are methodologies developed out of this process of reconstructing social influence, but he also does not directly tackle the difficult epistemological issues that confront evaluative critics of science.[8] And, as Eagleton and Longino have noted, his later conception of knowledge/power is a blunt instrument rather than a useful analytic device (Eagleton, 1991, 7–8; Longino, 1990, 204).

Foucault was one of the first to articulate a clear social constructionist position with regard to science, and through the 1960s, '70s, and '80s his work progressed in the sociology of science toward a stronger relativist position. In the past three decades, the field of science studies has proliferated tremendously and includes feminists,[9] philosophers, historians, social scientists, and natural scientists. All have done work that generally can be considered part of the paradigmatic shift toward conceptualizing the social construction of science knowledge.[10] Space and scope do not permit me to completely cover all this work here, but I will highlight some of the major works and ideas that have emerged and cover a few selected theorists in depth.

Bloor (1976) and Barnes (1974) challenged the heart of science legitimacy by developing social constructionist accounts of the development of mathematical theories. Bloor termed this work the "strong program" in the sociology of science. Latour and Woolgar (1979)did excellent micro-studies in which they followed the development of facts in laboratory life. Lynch (1985)[11] has done similar work at the micro-level, looking at how science fact and theory are formulated. Bloor (1976) has a more conservative epistemology, with his strong program formulation, than Latour and Woolgar (1979), but none of them seems to show how their work can be used to evaluate competing claims. Bloor's work has allowed for a space for good science, or at least experimental work, in his studies. Latour and Woolgar have understood the experimental site to be mediated by technology that is in itself socially constructed. It seems more difficult in their work to delineate any notion of "good science." This same tension follows other work in the history and sociology of sciences.

At this point, the various epistemological positions fall on a continuum that ranges from neopositivist to radical relativist with no one position dominant. The "neo-Popperians" hold that the sociocultural context does influence the formation of facts and theories, but the processes of observation, experimentation, replication, publication, and review serve to separate the "good" from the "bad" and, in the end, science still produces accurate representations of the world, if not truth. This position begins from a belief in the uniformity of reality and the possibility of objective knowledge. In this regard, it shares this belief with more traditional empiricist and positivist positions that begin from the belief that reality is limited, predictable, and knowledgeable. This position also accepts the influence of social factors on the production of science knowledge but sees this as a partial influence that is corrected by the proper use of method and established practices and by feedback from attempts at applying the theory to technological development. Some, such as Richards (1987), see this as a kind of evolutionary process in which the method of science ultimately "selects" the best ideas to be promoted to the status of accepted theories. These influences can be exerted on the metaphors used to think about problems and on the theories used to develop experiments and, ultimately, the experimental facts of a discipline. For example, Darwin may have gotten his idea for natural selection and survival of the fittest from Malthus, and he, in turn, may have gotten it while considering the social problems thrown up by a newly urbanizing capitalism. Its "truth" or "falseness," however, was determined in its vetting as a would-be scientific concept. This would be the point at which the rigors of method and practice would be brought to bear, which would determine its ultimate acceptance or rejection. This is the most "conservative" of the social constructionist positions.

On the other end of the continuum are those who carry epistemological relativism further and they are the "strong" constructivists. This group "questions not merely the autonomy but the epistemological integrity of science" (Longino, 1990, 9–10). Constructionist research employs very different conceptions of epistemology and the nature of science and science knowledge. Constructionists, as a rule, do not believe in more traditional positivist ideas about the separation of subject from detached observer and about the possibility of producing objective truths detached from specific social contexts. The most radical position begins from the belief that reality is too complex, interrelated, and unbounded to be known and ordered and that our preexisting conceptions influence what we assume reality to be. Ultimately, we find the "reality" that we look for, and no version of reality is any more correct than any other version. Technological success is no guarantee of truth. Getting something to work is not identical to understanding how and why it works. This position can be seen in Aronowitz's (1988) *Science as Power*, in which the author asks the following questions:

> What do we know when we know something? If our knowledge is ineluctably bound up with the processes of observation and experiment, and if these are permeated with *a priori* assumptions about the character of the observed or the uses to which knowledge is destined to be put, in what sense can we say that science as a type of knowledge is superior to metaphysics, including religion? (17–18)

He appears to answer these questions later when he outlines the book's key arguments:

> I argue that science is a labor process like others; that its practices constitute an intervention of a specific kind, whose contrast with other types of social and natural interventions *cannot be arranged hierarchically on a scale of truth or adequacy* [author's emphasis]; and finally that science is a discourse that narrates the world in a special way. (34)

Relativists relinquish all hope of providing any epistemological Archimedean point on which to stand when developing evaluative critiques. As I indicated above, one of the weaknesses of this position for critics of human sociobiology is that it does not appear to allow one to evaluate competing knowledge claims. Positions such as this have enraged many scientists, have started the science wars, and have produced a number of counter-attacks and counter-counter-attacks (Gross and Levitt, 1994; Gross, Levitt, and Lewis, 1996; Ross, 1996; Segerstråle, 2000b).

There are some, however, working in science studies who seek a middle ground. They seek to understand how things such as power, self-interest, and larger ideological discourses influence and demarcate what can count as science fact and theory in any given field. At the same time, many want, however, to maintain some special place for the kind of work that scientists do and for the integrity of the core of good science that includes debate, review, skepticism, and strong epistemic values. In the next section, I examine some of these positions.

Standpoint Theory and the Creation of a Modest Epistemological Middle Ground

Moderate constructionists do still hold out for some privileged place for science knowledge in the continuum of human discourse and at the same time insist on the inevitability of the always socially overdetermined character of all science knowledge. Theorists go at this middle ground in a variety of ways. This is a position that accepts the belief in the fixity and partial knowability of reality, while granting more importance to the social dimension than do the neopositivists. In this scenario, knowledge is always socially produced and only partial and fragile. Social factors permanently bind science knowledge, making it impossible to produce truth, but making it possible, with proper method and practice, to develop a coherent, accurate, and useful picture of the world. The key dimension to this critique is to explore and uncover the ways in which the social world helps or hinders this process in specific cases.

Haraway (1989) gives an interesting take on walking this middle ground in the Introduction to *Primate Visions*. She talks about four temptations that emerge when analyzing scientific discourse. One is the temptation to reject any possibility of realism and adopt a "thoroughly social and constructionist position." The second temptation is to adopt the Marxist tradition,

> which argues for the historical superiority of particular structured standpoints for knowing the social world, and possibly the "natural" world as well.... The tradition indebted to Marxist epistemology can account for the greater adequacy of some ways of knowing and can show that race, sex, and class fundamentally determine the most intimate details of knowledge and practice, especially where the appearance is of neutrality and universality. (Haraway, 1989, 6–7)

The third temptation is that of seeing and accepting what is done by scientists as practice that does "get at" the way the world really works. The fourth temptation "is to look always through the lenses ground on the stones of the complex histories of gender and race in the construction of modern science around the globe" (Haraway, 1989, 8). Haraway recommends employing all four perspectives "without succumbing completely to any of the four," lest one position may finally "[silence] all the others" (6). My own approach is somewhat similar.

I have found a number of positions useful and I delineate them here. For Hull (1988), a philosopher of science, "one of the tenets of the 'new' philosophy of science is that there are no 'raw data,' no theory neutral atomic facts" (19). This being the case, some such as Hull have chosen to redefine what we mean by objectivity:

> Quite obviously science is a social process, but it is also "social" in a more significant sense. The objectivity that matters so much in science is not primarily a characteristic of individual scientists but of scientific communities. Scientists rarely refute their own hypothesis, especially after they have appeared in print, but that is all right. Their fellow scientists will be happy to expose these hypotheses to severe testing.... The mechanism that has evolved in science that is responsible for its unbelievable success may not be all that "rational," but it is effective, and it has the same effect that advocates of science as a totally rational enterprise prefer. (4)

Longino (1990), a feminist philosopher of science, refurbishes the notion of objectivity with what she calls "contextual empiricism." As with Hull, she states that objectivity does not mean something that is possessed by individual observers. It is, instead, "the subjection of hypotheses and theories to multi-vocal criticism that makes objectivity possible" (213). This is, again, not an individual objectivity but a collective objectivity. It is the societal construction of objectivity rather than its quest by individuals. More recently, Longino advises that we develop epistemological humility and a "modest epistemology":

> An epistemology—as a theory of human knowledge—ought not to promise complete knowledge (or trade in other absolutes, like certainty) but ought to give sense to the distinctions and normative judgments that are a part of epistemic discourse. (Longino, 2000, 270–271)

Longino is influenced by other feminist standpoint theorists. These theorists have argued that societies create socially significant standpoints through which groups can create—in effect—different objects of study and can generate different kinds of science knowledge. Harding (1998) traces the development of standpoint theory back to Marx and Marxist notions of the "standpoint of the proletariat." She shows how it was reworked as a feminist epistemology most cogently by Dorothy Smith (1987, 1990, 1999), who starts off "research projects from issues arising in women's lives rather than from only the dominant androcentric conceptual frameworks of the disciplines and the larger social order" (Harding, 1998, 149).

From its beginnings in feminist research, this position has been taken up "in the knowledge projects of all the new social movements" (Harding, 1998, 149). What most standpoint theorists understand by this concept is important to grasp. Standpoints are not simply "cultural biases." According to Harding, "A standpoint is an object position in social relations as articulated through one or other theory or discourse" (150). People such as workers, feminists, gay rights advocates, and postcolonial researchers, scholars, and activists can use it as a position from which to make claims about power relations and their understanding of how society works from the standpoints of these groups constituted by social relations of power and inequality.

While standpoints can provide important counter-hegemonic insights, facts, and knowledge about the world, I do not understand standpoints to be *a priori* positions available only to those who occupy a particular social niche. Social locations do not provide ways of knowing in some essentialist fashion. Anyone can achieve the insights available through various standpoints. What standpoints offer, in D. Smith's sense, are foci and points from which to begin analysis. These are not the exclusive domains of groups but rather generate their uniqueness from their starting points, which can be adopted by willing researchers.

Equally important, standpoints are not reducible to individual viewpoints. They derive from structured positions of inequality in societies. As Harding puts it,

> a standpoint is not the same as a viewpoint or perspective, for it requires both science and political struggle, as Nancy Hartsock puts the point, to see beneath the surfaces of social life to the realities that structure it. (150)

Harding contends that standpoints should be understood as multi-dimensional projects, encompassing:

> a philosophy of knowledge, a philosophy of science, a sociology of knowledge, and a moral/political advocacy of the expansion of democratic rights to participate in making the social decisions that will affect one's life, and a proposed research method for the natural and social sciences. (161)

Standpoint theory is especially relevant to studying the sociobiology debate. Ultimately, alternative formulations will require that we take up the standpoint of organisms as well as the standpoint of genes and molecules in order to produce accounts that can accommodate the multidimensionality and multilayered reality that is life.

Standpoint theory also informs our understandings of what should count as objectivity and truth. In this context, Longino (2000) has chronicled the rise of what she sees as a modified pluralism within the philosophy of biology:

> Philosophers of biology who advocate pluralism, especially those advocating strong forms of pluralism, are claiming that whatever may be the case in physics, the complexity of organic and living entities and processes eludes complete representation by any single theoretical or investigative approach. Any given approach will be partial, and completeness will be approached not by a single integrated theory, but by a plurality of approaches that are partially overlapping, partially autonomous, and resisting reconciliation...[and] quantitative data will vary between approaches, so that comparing data descriptions will show inconstancies between the approaches. Deep insight into a phenomenon will be purchased at the cost of losing unifiability with other approaches. (Longino, 2000, 268)

She acknowledges that this position is in contradiction with our more established notions about the unity of the world and the unity of true or at least accurate statements about the world. Her advice is that we acknowledge that we cannot determine if this plurality of representations describes how the world is or just describes our only way of viewing this reality.

For example, Longino (1990) argues in a comparison of hormonal and selectionist models of human gender role behavior, that neither has more "evidential status" than the other.

> Both rest on explanatory models that involve metaphysical assumptions about causality and human action. Neither theoretical perspective can muster constitutively based arguments sufficient to exclude the other—thus both can continue to generate studies that are used to support microhypotheses about the etiology of particular forms of behavior that are consistent with one or the other broader model. (161)

In such cases, standpoints and interests become important determinates of the shape of research and the choice of model favored.

Modest Epistemologies and Scientific Pluralism

The ideas of multiple standpoints and a modest epistemology also call forth notions of pluralism in explanation and the co-existence of differing and possibly contradictory explanations. In such cases, Michael Ruse's (1999) formulations also are useful for those seeking to develop a middle ground in the empiricist-constructionist divide. He talks about two kinds of values in science. One he calls "epistemic" values. These values encompass the standard processes embodied in the scientific method such as experiment, reproducibility, publication, and peer scrutiny that one hopes guarantee some measure of truth. The second is what he calls the cultural or social values. He sees these two sets of values interacting and influencing the work of scientists in different ways at various points in the development of a specific theory.

The case he examines is the history of the development of evolutionary theory. His compromise involves an acceptance of the always-present cultural influences and the need to strive for strong epistemic values. As with Hull, Ruse does not hold out for truth per se. In quoting Hilary Putnam, Ruse seconds Putnam's idea that "truth" is some sort of "(idealized) rational acceptability—some sort of ideal coherence of our beliefs with each other and with our experiences *as those experiences are themselves represented in our belief system*—and not correspondent with mind-independent or discourse-independent 'states of affairs'" (cited in Ruse, 1999, 252). For Ruse, the context always influences the forms of science knowledge generated, "but some formulations are better than others," and success can be measured through assessment of general productivity and explanatory power. The epistemic values make science special but always bound by the cultural influences.

From this position, Ruse is able to judge Wilson's human sociobiology. He concludes that "as one turns to sociobiology, the non-epistemic factors start to rise, and conversely the epistemic constraints sometimes get looser. This is certainly the case when Wilson treats our own species" (Ruse, 1999, 192). He goes on to discuss Wilson's notion that religiosity is physiologically programmed and concludes: "At points like this, non-epistemic urges [read "cultural or personal biases"] are substituting for evidence conforming to strict epistemic norms" (192). Longino's and Ruse's conceptions are to me complementary. Longino's work clears the way for a modest epistemology and a pluralism of standpoints, while Ruse's work is prescriptive. It tells us what to look for and how to judge the knowledge claims of competing standpoints.

It may be, however, that "judgment" can only be rendered in a partial fashion. Kellert, Longino, and Waters (2006) hold that a difficult aspect to accept of this scientific pluralism is the implication that "scientific approaches and theories should not be evaluated against the ideal of providing the single complete and comprehensive truth about a domain" (xxiv). What is also difficult to accept is that this kind of scientific pluralism undermines the assumption that "the aim of science is to identify the single comprehensive truth" about a domain (ibid). They hold that, ultimately, scientists may have to come to accept that

> different descriptions and different approaches are sometimes beneficial because some descriptions offer better accounts of some aspects of a complex situation and other descriptions provide better accounts of other aspects. And this may be the way it will always be. (ibid)

And so, even in the debate over the influences on behavior, Longino points out that, while most people now hold that both nature and nurture are important, "most seem also to believe that only their approach successfully articulates the nature of the interaction and hence produces genuine knowledge" (Kellert, Longino, and Waters, 2006, 102).

In light of a modest epistemological position, Longino sums up the state of knowledge as partial, plural, and provisional (2002, 207). A number of assumptions follow from the acceptance of a middle or modest epistemological position regarding truth claims in science and the nature of science knowledge in general. First, there is a "core of good science" with established epistemic norms that stand as a guide and boundary for knowledge claims. In relation to the sociobiology debate, this position allows one to argue for Ruse's strong epistemic values in science practice and theory and to argue that there are clear instances in which deterministic human sociobiological proclamations swerve widely from this core. Second, there is always social influence at all levels and junctions in scientific practice. This social influence always shapes science claims and also the quality of epistemological supposition itself. What we can do is examine these influences as clearly as possible. In this regard, one can include research on the critiques of human sociobiology that show this to be the case, as in the examples I gave earlier in this chapter. Third, strong objectivity cannot be understood as disinterested and abstracted exploration. Rather, it is strongest in societies that work to foster a plurality of claims and standpoints. Fourth, these claims and standpoints, however, can produce incommensurate knowledge, and the best we can hope for is to reject the possibility of a strong and unitary positivism and embrace a modest epistemology in which claims are always partial, incomplete, and possibly contradictory.[12]

Wilsonian sociobiological discourse clearly promotes a neopositivist notion of science and epistemology. In this regard, critics must examine Wilsonian human sociobiology for the ways in which it tries to limit the inclusion of alternate standpoints in the debate. In the end, what we can work toward is knowledge that favors complexity over reduction and accords standpoints their "angular vision"[13] and accords levels of reality their uniqueness. This also must include working toward textbook presentations of science that favor complexity over reduction and that accord space for multiple and possibly contradictory standpoints—that is, textbook presentations that allow space for multiple presentations and for controversy. There is a theoretical space emerging within the science education literature that allows for these kinds of discussions within textbooks. The debate about the nature of science that has happened within science studies disciplines has also emerged at around the same time in the science education literature. Work in this literature has attempted to theorize why and how discussions of the nature of science can and should be presented in classrooms and in textbooks.

Related Theoretical Developments on the Nature of Science in Science Education

I have thus far, in this chapter, presented work by philosophers, sociologists, and historians in the general field of science studies. This work has moved toward a more socially contextualized understanding of the practices of scientists and the development of science knowledge. Likewise, many philosophers, sociologists, historians, and curriculum specialists who work in the general field of science education also have debated, and continue to debate, the epistemological status of science knowledge and have begun to articulate less positivist conceptions of science and science claims. Like their counterparts in science studies, these researchers also have come to see the importance of both clarifying epistemological issues and including the current range of debate about the epistemological status of science knowledge in the teaching of science teachers and science students. This is a body of work that also challenges the idea of a decontextualized science practice and the idea of absolute truth. It insists on locating any given claims within the sociocultural and political climate in which they appear and prosper (Hodson 1994; H. Rose, 1994b; Taylor and Cobern, 1998). This work includes the constructivist movement (Cobern, 1998a, 1998b). It also includes what began as the Science Technology and Society (STS) movement (Aikenhead, 1994a, 1994b; J. Solomon, 1993, 1994a, 1994b; J. Solomon and Aikenhead, 1994), which has evolved into the Science, Technology Society, and the Environment (STS(E)) movement (Aikenhead, 1989; Hungerford, Bluhm, Volk and Ramsey, 1998) and the nature of science movement in science education (Matthews, 1994, 1998a; McComas, Clough and Almazroa, 1998).

Constructivism

In a fashion similar to the effect constructionism has had in science studies, the movement away from positivism has been ignited and reinforced by constructivism in science education[14] (Matthews, 1994). Also, there is often direct cross-fertilization and cross-referencing between science studies and science education writings (Slezak, 1994; Suchting, 1997). In the same way that there are many versions of constructionism in science studies, there are many versions of constructivism in science education. An epistemological continuum of constructivisms has emerged in science education similar to the coninuum I discussed above in the context of constructionist epistemologies in science studies (Geelan, 1997; Loving, 1997).

There are radical constructivisms (von Glasersfeld, 1989, 1992) and more modest approaches (Duschl, 1990; Suchting, 1997; Bickhard, 1997) with numerous critiques and debates from all quarters concerning the usefulness or harm that constructivisms encourage (von Glasersfeld, 1992; Suchting, 1992; Nola, 1997; Phillips, 1997; Grandy, 1997; Kragh, 1998). There also has been a kind of "science war" as well between the neopositivists (Slezak, 1998) and constructivists (Suchting, 1997). The cumulative effect of the constructivist movement on science education has been very strong and is conveyed by this statement from Cobern: "[M]any educators have come to see science as one of several aspects *of* culture.... It follows that science *education* is an aspect of culture" [author's emphasis] (1998a, 7).

Constructivism in science education also has fostered the development of a movement toward more critical and evaluative science teaching. Indeed, in much the same way that I call for the development of a critical sociology of science and attempt to begin to evaluate the sociobiology debate, there are voices in science education calling for the development of a critical science education (Hodson, 1994; Taylor and Cobern, 1998). This emergence of an international movement advocating the development of a critical edge or focus to science teaching is exemplified in *STS Education: International Perspectives on Reform* (Solomon and Aikenhead, 1994) and in *Socio-Cultural Perspectives on Science Education: An International Dialogue* (Cobern, 1998b).

STS(E) and Its Predecessor, STS

This movement toward a critical scrutiny of the claims and effects of modern science and science knowledge also has been encouraged and fostered initially by the Science Technology Society (STS) movement and then by its successor, the Science Technology Society and Environment (STS(E)) movement within science education (Aikenhead, 1989, 1994a, 1994b; Hungerford, Bluhm, Volk, and Ramsey, 1998; Solomon, 1993, 1994a, 1994b; H. Rose, 1994). According to Aikenhead, the Science, Technology and Society movement has linked three elements in science education. He describes the linkages in this way:

> The study of the natural world we call science. The study of the artificially constructed world is technology. And society is the social milieu. Teaching science through science-technology-society (STS) refers to teaching about natural phenomena in a manner that embeds science in the technological and social environments of the student.... In other words, STS instruction aims to help students make sense out of their

> everyday experience, and does so in ways that support students' natural tendency to integrate their personal understandings of their social, technological, and natural environments.
>
> STS is also expected to fill a critical void in the traditional curriculum—the social responsibility in collective decision making on issues related to science and technology. (Aikenhead, 1994a, 48–9)

The science-technology-society-environment (STS(E)) movement adds to this orientation a focus on the impact that the social, scientific, and technological realms have on the quality of our environments. It encourages students to take responsibility with the aim of understanding and controlling human behavior, and with the goal of preserving and improving environmental conditions. STS(E) seeks to locate science practice and science knowledge in the social contexts in which they arise and to examine their impact on this social world. This movement to contextualize and politicize science practice and science knowledge has met with the same kinds of resistance that have sparked the science wars in science studies (Solomon, 1994a).

The Nature of Science

The debate and controversy sparked within science education by the constructivist and STS(E) movements have occurred parallel to the movement in science education to make consideration of the nature of science itself a fundamental part of science teaching. Considerations of epistemology and sociology in relation to science education have a history (Cawthron and Rowell, 1978). According to Matthews (1998a) "there has been a long tradition of theoretical writing on the subject of establishing the cultural, educational, and scientific benefits of teaching about the nature of science" (xi). He traces the roots of this writing back to Ernst Mach in the late 19th century and John Dewey in the early twentieth century (Matthews, 1998a). McComas, Clough, and Almazroa (1998) define the modern movement in the following way:

> The nature of science is a fertile hybrid arena which blends aspects of various social studies of science including the history, sociology, and philosophy of science combined with research from the cognitive sciences such as psychology into a rich description of what science is, how it works, how scientists operate as a social group and how society itself both directs and reacts to scientific endeavors. (McComas et al., 1998, 4)

As with the constructivist and STS(E) movements, there are those who fear that contextualizing and politicizing science practice and knowledge will have a detrimental effect on science teaching and curricula (Lederman, 1995). Owing to the complexity of the issues involved in presenting the nature of science to students, McComas et al. (1998) argue that "where consensus does not exist, teachers should present a plurality of views" on the nature of science (6). Indeed, the movement to elucidate the nature of science in science education has potentially radical and profound implications that can transform our traditional conceptions of the status of science practice and knowledge. McComas (1998a, 53–68) illustrates this radical potential in his elucidation of the ways in which his interpretation of the nature of science dispels a series of myths about science prevalent in traditional science education. These myths are:

Myth 1: Hypotheses become theories that in turn become laws

Myth 2: Scientific laws and other such ideas are absolute

Myth 3: A hypothesis is an educated guess

Myth 4: A general and universal scientific method exists

Myth 5: Evidence accumulated carefully will result in sure knowledge

Myth 6: Science and its method provide absolute proof

Myth 7: Science is procedural more than creative

Myth 8: Science and its methods can answer all questions

Myth 9: Scientists are particularly objective

Myth 10: Experiments are the principal route to scientific knowledge

Myth 11: Scientific conclusions are reviewed for accuracy

Myth 12: Acceptance of new scientific knowledge is straightforward

Myth 13: Science models represent reality

Myth 14: Science and technology are identical

Myth 15: Science is a solitary pursuit

The parallels in work done on the nature of science in science studies and science education are important. According to Longino, in light of the partial, plural, and provisional nature of knowledge, the challenge is "not so much a metaphysical one but an educational one" (Longino, 2002, 202). In this regard, Longino asks the question: "How can the value of scientific research as a source of guidance for policy decisions be maintained in the face of the complexity of nature and the partiality and plurality of our knowledge of it?" (ibid). Social spaces and arbiters or repositories of the various scientific pluralities need to be created so that "objectivity" at the social level is nurtured and maintained (Longino, 2002; 1990). Textbooks can be important sites where competing claims are assessed and compared and where pluralities of explanations are allowed both theoretical and social space in which to thrive. Textbooks can act as conduits that link students with a pluralism of accounts and explanations in specific realms. To do this, authors must enlarge their understandings and presentations on the nature of science. This would mean presenting not just a positivist view of science but also more modest epistemological positions. It means providing social context to discovery and controversy in general and providing sites where pluralities of explanations are not just mentioned but fully presented. One can ask if issues of epistemology and the nature of science are discussed. One can ask if there is an attempt in the text to be pluralistic only in form, that is only by "mentioning" things such as environmental and cultural influences; or whether the attempt at pluralism has depth and breadth. These are some of the themes that are explored in my analysis of specific textbook presentations of the nature of science in Part II of this book.

Notes

1. Both Levins and Lewontin have continued developing the application of these principles in their later books. Haila and Levins (1992) have co-authored *Humanity and Nature: Ecology, Science and Society* and Lewontin (2000b) has written *The Triple Helix: Gene, Organism, and Environment*.
2. McKinnon holds that proponents of evolutionary psychology ignore evidence, especially in anthropology, that contradicts their positions. She also holds that, at their philosophical foundations, evolutionary psychologists have "presupposed the natural universality of their own categories and understandings" (11–12) such that "a self-consciously Victorian morality of sex, gender, and family relations [have] been united with a neo-liberal economic ideology" (12) and created what McKinnon calls "neo-liberal genetics."
3. In his defense, Segerstråle (2000a) contends that Wilson's opposition to Marxism is really an opposition to his opponents' "political and activist side" (287). She also contends that he is "genuinely interested in the philosophical side of Marxism and even once told [her] in an interview that he read *Science and Nature*, a self-consciously Marxist scientific journal (whose first issue appeared in 1979)" (286).
4. Critics generally have in mind here the major social and political ideologies such as those that Marchak (1988) terms liberalism and conservatism or that Macpherson (1964) calls "possessive individualism."

5. For work on evolutionary psychology, see McKinnon (2005); and for work on genetics, see Paul (1998); G. Allen (1983, 1991, 1994); Hubbard and Wald (1993); and E. Keller (1992a, 1992b).
6. Segerstråle (2000a) argues that critics such as Lewontin seem to imply confused and confusing versions of what is good scientific method, at one time arguing against the process of reducing wholes to parts for analysis and at other times endorsing this same "reductive" process.
7. The key works in the pre-war period were Bachelard's *The New Scientific Spirit* (1984), Burtt's *The Metaphysical Foundations of Modern Science* (1932), and Bernal's *The Social Function of Science* (1939).
8. See also Dreyfus and Rabinow (1982); Gutting (1989); Hoy (1986).
9. Feminist sociologists, scientists, and philosophers of science have in many respects led the epistemology debate in science studies and made key contributions. See also Birke and Hubbard (1995); Garry and Pearsall (1989, 1996); Harding (1986, 1991, 1998); Harding and Hintikka (1983); E. Keller (1985, 1992a, 1992b, 1995, 2000); Longino (1990, 2000); H. Rose (1994); D. Smith (1987, 1990, 1999); and Tuana (1989).
10. There is a tremendous range of work in this field. Good overviews of this progress can be seen in Oldroyd (1986) and Shapin (1995).
11. See also Lynch (1991, 1992, 2001).
12. While philosophers of science may take exception to the comparison, Longino and Ruse appear to arrive at a similar place to Kitcher (2001) in his call for an epistemologically modest realism as a way of avoiding the polarizations developed during the "science wars" (11–28).
13. I borrow this concept from Wilfred Desan's *The Planetary Man* (1961, 1972). I use it in a very different sense than he did, however. For Desan, individuals each carry an angular vision, and his sense of this vision has decidedly theological overtones. For me, angular visions are views of the world generated by and through socially significant and mediated standpoints, not individual perspectives.
14. There is some debate over when to use "constructionist" and "constructivist." A thorough treatment of this issue is beyond this book. I am here adopting Erminia Pedretti's suggestion (as given in private conversation) that the term "constructionism" be used to describe the sociological notion of the social construction of the social and scientific worlds and that "constructivism" be used to describe the movement within education that for some is similar to what sociologists mean when they use "constructionism." There are, however, constructivisms that also can be more Piagetian, in the sense that people actively construct knowledge-in-action and that the frameworks people bring to learning shape how they understand what is taught to them (Matthews, 1997). Also, these frameworks may be impervious to teachers' transformative attempts. This is a not entirely satisfactory delineation, because I feel that some in science studies use the terms interchangeably while some in education use the term in a manner identical to some of the meanings given to constructionism (see, for example, Matthews, 1997).

PART II

The Textbook Study

· 4 ·

METHODOLOGICAL ISSUES AND TEXTBOOK SELECTION

Part II of this work is a qualitative textual analysis of selected introductory biology textbooks, in which the initial themes that are examined in the textbooks emerged from my analysis of the key issues in the sociology debate. This examination and analysis have been shaped and influenced by my analysis of the ways in which sociobiological discourse constructs its legitimacy, the obstacles that critics faced in attempting to undermine that legitimacy, and the many theoretical fronts involved in this contested terrain. My aim in the next five chapters is to begin to explore the ways in which the sociobiology debates and related issues as outlined in the previous chapters have filtered into a selection of biology textbooks.

Summary of Part I

My research on the last twenty-five years of debate over what can be called the "discourse of Wilsonian Sociobiology" has revealed a number of key conceptual positions or themes that bolster its credibility and legitimacy. Sociobiological discourse relies on a neopositivist epistemology that does not challenge the ability of science to make valid truth claims. In conjunction, it relies on a reductive or atomistic method that sees the taking apart or reduction of complex human behavior into "general" and "abstract" components as a potentially useful method to apply to any nonhuman and human behaviors. It also relies on a simple Mendelianism and a vague and undefined conception of "traits" to connect genes to complex behaviors. Last, it relies on a strongly adaptationist interpretation of evolutionary theory in its linking of genes, evolutionary adaptation, and complex human behaviors. Consequently, the controversial questions that emerged from the sociobiology debate centered on a number of questions: 1) What will constitute proper scientific method? 2) What are the key principles in genetics and evolutionary theory? 3) What can count as a biological trait? 4) What constitutes the nature of organisms? and 5) What claims can be made about the "truth" of scientific knowledge and the nature of science itself?

In concert with these foundational principles, this discourse of Wilsonian human sociobiology also relies on a number of rhetorical devices, which I discussed in Chapter 2, to frame its legitimacy. At the same time, this discourse maintains the boundaries of debate by leaving out substantive critiques and alternative models developed by opponents. These include alternative conceptions of organisms, more complex conceptions of both genetics and evolutionary theory, constructionist perspectives on the nature of science knowledge, and social theory that addresses the many levels of an organism's reality beyond the genetic and beyond the biological.

The Themes and Subthemes That Guide This Textual Analysis

The themes that emerged in my analysis of the ways in which the Wilsonian discourse of sociobiology structures its credibility and legitimacy inform the investigative beginning for my analysis of the textbooks. I have constructed the presentation of my research along four major themes or focal points. These are 1) presentations of sociobiology and animal behavior, 2) presentations of evolutionary theory, 3) presentations of genetic theory, and 4) discussions of the nature of science. Each of these topics is examined for each text analyzed. In Appendix Two, I discuss issues I faced in doing a textual analysis as qualitative research. These themes are all interrelated because sociobiological discourse relies on a simple Mendelianism, an ultra-adaptationism, a neopositivist conception of the nature of science, and various rhetorical devices. For example, in genetics it is crucial to look at the ways in which traits are presented and the ways in which genotypes and phenotypes are related. Also, it is important to examine how the general relationship between genes, cells, organisms, and behaviors is presented, both in consideration of genetics and in consideration of animal behavior and sociobiology. And it is crucial to examine the roles that genes have in general: Do the texts present a discourse of gene action uncritically or with qualifications? In evolutionary theory, it is important to look at the ways in which adaptationism is presented in relation to other dynamics that go toward creating a more complex understanding of the processes of evolution. All of these conceptions are controversial and debated within biology. In this analysis I look for ways in which the texts bolster, challenge, or ignore these issues and, in so doing, bolster, challenge, or ignore sociobiological and related discourses.

Within these textual analyses, I also look at a number of subthemes that arise out of my study of the sociobiology debate. These include an examination of rhetorical devices that I covered in Chapter 2 that Wilsonian sociobiological discourse employs, such as anthropomorphism, doublespeak, appeals to common sense, "propaganda of the middle ground," the elevation of specificity to the level of general principles, and reification. This also includes an examination of the process of "mentioning" that arises from both the study of the sociobiology debate and the textbooks themselves. In both instances, "mentioning" is the process whereby a controversy is dealt with by providing evidence to support one side and then simply "mentioning" the issues on the other side. It is a presentation that is fundamentally uneven and generally unhelpful to the reader.

In addition to the sensitizing themes that have emerged from my study of the sociobiology debate, two subthemes also emerged from my analysis of the textbooks themselves. One of the subthemes has been the ways in which inserts and pictures are used to highlight and separate information in the texts. I have noted this use of alternative presentation styles where I have understood them to be crucial to my general analysis.

A second subtheme relates to the presentation of the nature of science in the textbooks through the portrayals of important historical figures in science, such as Mendel and Darwin. Historical figures can be presented as located within a specific place and time. This type of presentation provides a social and political context for their work. It "constructs" their knowledge within this context as bound and limited by their context. Such a contextual presentation would be more compatible with a modest epistemology. The other way to present such historical figures and their work is to present them in a transhistorical fashion. In the presentation their insights and abilities stand apart from their social and political context and the knowledge they generate can be presented as "above" social influence. Such knowledge can be portrayed as "truth" and the scientists given the status of canonical figures.

In this regard, presentations of Mendel and his work are particularly interesting because of controversies that emerged around his work. Mendel is usually cited as the father of modern genetics, and his experiments on plant hybridization are cited as the model of proper scientific method and practice that yielded neopositivist "facts" about the biological world. However, both the likely veracity of his reported results and his fatherhood of the discipline have come into question. Brannigan (1981) cites Sir Ronald Fisher's (1936) conclusion that "Mendel's ratios were far too accurate to have occurred by chance" (Brannigan, 1981, 102). Fisher has surmised that Mendel probably "had his theory of hybridization worked out before he began the painstaking and extensive breeding experiments" (cited in Hubbard, 1982, 67; in Rose and Rose, 1982). Brannigan also cites Zirkle's (1951) study of Mendel that claims that work published by Jan Dzieron eleven years prior to Mendel's already had detailed the precise hybrid segregation ratios Mendel is credited with "discovering" (Brannigan, 1981, 102–106).

Brannigan goes further to assert that Mendel's work was never rediscovered; it "was not revived in 1900, but constructed then for the first time" (Brannigan, 1981, 103). Brannigan also contends that Mendel's work in its discovery did not represent a radical break with other plant breeders' work, but, rather, was the same kind of hybridizing work. It was the context of the rediscoverers that forced a reinterpretation of Mendel and a reification of his work into what have come to be called "The Laws of Inheritance."

> The status of Mendel's work changed over time with a shift in contexts from a concern with hybridizing as a key to evolution pure and simple, to a concern with hybridizing as a clue to a theory of inheritance which would complement the theory of natural selection. Hence, the status of discovery was not a simple function of the *contents* of the paper as much as the *contexts* in which it appeared. (Brannigan, 1981, 89)

Even though this interpretation of the "rediscovery" of Mendel and the assessment of the veracity of Mendel's experiments are subject to debate (Sapp, 1990b), it is possible to present Mendel's work as an example of the ways in which investigators' assumptions and social contexts shape how fact and theory emerge and are treated. Texts can either present more neopositivist, canonical, and reified presentations of Mendel and his work, or more constructivist and epistemologically modest presentations. The choice has implications for students' conceptions of genes and gene action as much as it does for students' conceptions of the nature of science. The sociobiological discourse focuses on Mendel's laws and uses them to legitimate its reductionistic connections between biology and behavior. Tempering these "laws" reduces their general applicability and the general ability to "trait fix." I had not considered either of these issues—the use of inserts or the treatment of historical figures—before beginning my examination of the texts. These subthemes emerged as a product of the textual analysis. As a result, I examine how figures such as Mendel and Darwin are presented in the texts—in the contexts of the nature of science debates and in the contexts of their home disciplines.

These data cannot be presented in a formulaic manner. It is difficult to compare texts on each point because they differ in what they cover and in the contexts from which issues emerge. What I have tried to do instead is weave a narrative for each text that I hope captures the overall impressions and emphases of the texts in the contexts in which the issues arise. In general, I have found in the textbooks and in the debates a very tight correlation between how one thinks about human sociobiology and how one thinks about genetics, evolutionary theory, and epistemological issues raised in the nature of science controversy. The more determinist, simplistic, and canonical the emphases in presentations of genetics and evolutionary theory, the more likely I was to find prosociobiological formulations and uncritical neopositivist assumptions about the nature of science itself.

Why Textbooks?

Science textbooks are probably the most important resources in the teaching and learning of science (National Academy of Science, 1997, cited in Pozzer-Ardenghi and Roth, 2004). There is not, however, a lot of research available on "how textbooks are used by teachers and students" (Driscoll, Moallem, Dick, and Kirby, 1994, 80). Driscoll et al. suggest that many of the studies that have been done tend to assume that textbooks are the major influence on instruction (80). In their study of textbooks' contributions to learning in a science class, Driscoll et al. found that teachers tended to use the textbook as a reference source for canonical information. Gottfried and Kyle (1992) found that the majority of the teachers they studied "function as passive, uncritical technicians" in relation to textbook use and to biological fact and theory in general (46). If science textbooks serve as canonical reference sources for science teachers, then biology textbooks should anchor not simply a specific paradigm in biology as much as the practices and processes of science that generate accurate and reliable conceptions of reality. It is crucial that these texts present a full range of debate on issues as controversial and influential as human sociobiology and the general topic of biology and human behavior.

Conceptualizations of the relationship between biology and human behavior will become increasingly important as biotechnology advances, and these conceptualizations will continue to be controversial. In the "modern" world, the bulwark of certainty has surrounded our conceptions of knowledge, but this has given way to a postmodern world with an abundance of unanchored and often directly manipulated information. In this postmodern world, the special status of science knowledge has been undermined, and the whole notion of truth and veracity is under attack (Miller, 2008). Textbooks become increasingly important anchors in this situation. Textbooks should be important sites where controversies in science can be presented so that students, teachers, and the public can develop skills to critically judge competing theories. Likewise, science textbooks should also be sites where issues around the nature of science are debated. Similarly, biology textbooks should be places where controversies about the relationship between biology and behavior are presented. Curriculum content in general and textbook presentations in particular are active opportunities for what Manley-Casimir has called "contested curricula."[1] These textbooks structure legitimacy for positions and should help foster the development of informed opinion on scientific controversies, such as conceptions of science, science knowledge, and nature.

Textbooks also present an interesting boundary study for analyzing the movement of new paradigms from the "laboratory" to the general public.[2] Textbooks are important as well because, more and more, they tend to be cooperative documents, usually involving consensus and consultation within segments of the scientific community. In theory, they should reflect a range of opinion on current debate within the field. Analysis of the status of the human sociobiology project within academe can be done, to some extent, by studying the ways in which this debate and the related legitimating positions are presented to teachers and students in biology textbooks. Likewise, analysis of textbooks gives some indication of the ways students are being trained to think about controversial issues in science.

I have found little evidence in the literature that there have been many systematic studies of textbook presentations on sociobiology, let alone on the controversy surrounding it. Herzog (1986) looked at the presentations of sociobiology in psychology textbooks. Spencer (1991), in an unpublished study, did look at the ways in which sociobiology was covered in general biology textbooks. He concluded that, while most textbooks in the time frame he chose to examine included material on sociobiology, the texts showed no general pattern of coverage and that "there were not many commonalities amongst words defined in the glossaries, text coverage and the indexing of material."

Which Textbooks? The Selection of the Texts

The selection of textbooks to analyze proved to be a complicated task. This textbook analysis has been done in two stages. I first began to look at introductory biology textbooks in 1999. At that time, Ontario high schools went from grades 9 to 13 and Ontario Academic Courses (OAC) included a range of subjects such as biology, chemistry, physics, and philosophy that were offered to grade 13 students. The content of the OAC biology courses was similar—and in some cases identical—to introductory biology courses offered at universities in Canada and the United States, high-school advanced-placement courses offered in the United States, and International Baccalaureate (IB)[3] program courses.

The Ministry of Education of the Province of Ontario maintained a list of approved textbooks from which high schools could select the texts to be used in a given OAC course in a given school. This list was called Circular 14[4] (see Appendix One for the complete list). I decided to confine myself initially to three texts chosen from this list. At the time, however, the provincial government was in the process of eliminating grade 13 and so had stopped reviewing or adding to Circular 14 in 1997. Books on the list were granted interim approval until new curriculum guidelines were created, so my initial sample was taken from books published before 1998. I decided to use the grade 13 texts because I had a defined and approved list from which to select and because these texts would be more advanced than basic biology course texts and more likely to cover the sociobiology question. In general, I did find that lower-level texts did not include a chapter on animal behavior and thus did not cover the specific topic of sociobiology. [5]

I used three criteria for this selection of the four textbooks I chose. The first criterion was popularity.[6] I wanted books that were widely used in southern Ontario. I did not have the resources to create such a database[7], and because of the in-depth nature of this textual analysis I could not include every book available because I examined approximately 700 to 800 pages of text in each book. The second criterion was that I also wanted, if possible, a Canadian text.[8] My third criterion was that I wanted to choose texts that display a range of content differences in the presentation, both of sociobiology and the related issues that I outlined above. In this way, my sample can be seen as very purposive and it is without doubt influenced by my concerns and biases. However, I feel that I have gone to great lengths to ensure that a wide spectrum of presentations is encompassed in these texts. Ultimately, I am doing an analysis of general positions in a spectrum as much as I am engaged in an analysis of specific texts.

I reviewed all the books on the circular and talked to administrators, teachers, and consultants at various schools, boards, and districts.[9] Through these conversations it seemed that two texts stood out as the most popular. One was Galbraith's *Biology: Principles, Patterns, and Processes*,[10] a Canadian text; and the other was Raven and Johnson's *Understanding Biology* (3rd edition).[11] I estimate from these informal fact-gathering conversations that these two texts were likely used in 50 to 60 percent of OAC biology courses in southern Ontario.[12] Two other books stood out because they presented the subject matter in unique and important ways.

The third book I chose was Levine and Miller's (1994) *Biology: Discovering Life* (2nd edition). While not as widely used, it took up more controversial issues than most other texts. It appears atypical in that it seems to engage controversial issues in biology more directly than most texts. Therefore, it could provide a unique and an important perspective when considering so controversial a topic as human sociobiology.[13]

The fourth book I chose was *Biology* (4th edition) by Neil Campbell (1996). The Campbell text appeared to enjoy widespread popularity in high-school advanced studies programs, International Baccalaureate (IB) Programs, and introductory biology courses at the university level in Canada and the United States. The Campbell textbook also is interesting because it intersperses interviews with biologists throughout the text, and the fourth edition included an interview with E. O. Wilson. I have used the fourth edition of *Biology* by Neil Campbell even though Circular 14 included only the third edition (1993). This is because the updating of Circular 14 ended before the fourth edition could be added. I chose the fourth edition not simply because of the interview with Wilson. I chose it because the third edition was published in 1993 and it was older than the other books I reviewed. The fourth edition, which was published in 1996, offered a more effective comparison because its publication date was closer to the other books reviewed in this phase of the research.

In 2009, I developed a second phase of this research when I decided to go back and look at newer editions of the same books that I had chosen for Phase I. This longitudinal perspective is important in that it would allow me to track changes in the authors' approaches to the themes I examined, which in turn would allow me also to gauge the current state of biological thinking on these issues. I found that only two books had continued to be updated and published at the same level of sophistication. These are *Biology* (8th edition), written by Losos, Mason, and Singer (2008), based on the work of Raven and Johnson and published by McGraw-Hill Higher Education; and *Biology* (7th edition), written by Campbell and Reece (2005) and published by Pearson Educational as Benjamin Cummings. In both cases, new authors had joined the writing teams. The analyses of the more recent editions of these two texts are covered in Chapters 5 and 8. The other two sets of authors whose textbooks I had reviewed in the first phase of this research published texts after 1997, but these are written for lower-level biology courses and so do not have the depth of information to make them relevant to my analysis.[14] Ultimately, my choices have meant that this analysis is a commentary on texts used in advanced placement courses in Canadian and U.S. high schools and in introductory courses in university programs.

Notes

1. Michael Manley-Casimir, 1999; private conversation.
2. For a more in-depth discussion of boundary work in science studies, see Gieryn, *Cultural Boundaries of Science*, 1999.
3. The IB program is an advanced academic high-school program offered all over the world. Examination structures and evaluations are standardized worldwide. If students in the IB score at a particular level, wherever in the world they participate in the program, they are very likely to receive introductory biology credit at the university they attend. Therefore, this text has international relevance as well as relevance for university-level comparison.
4. After the structural reorganization of Ontario high schools, this list was renamed the Trillium List.
5. When Grade 13 was phased out in Ontario, the OAC program in biology was replaced with a grade 12 senior biology program. In the case of senior-level science courses, these new curriculum documents were released in late 2000, and no new textbooks were being added before then. In informal talks with publishers' representatives at the Science Teachers Association of Ontario annual meetings in Toronto in November 1999, I found that publishers were waiting to see the new curriculum documents before beginning to produce texts for the new senior biology programs. When I began my research and analysis of the texts in late 1999 and early 2000, there were as yet no senior-biology curriculum guidelines and no indication of new textbooks from publishers. The guidelines were not released until the summer of 2001.

   Also, the Ontario government has mandated that the new grade 12 senior biology course, which was implemented in the new curriculum and which replaced the Grade 13 OAC biology course, must use made-in-Canada textbooks by Canadian authors. Therefore, it was no longer possible to simply import textbooks from the United States or elsewhere for this new senior biology course. These and future editions, however, likely will continue to be used as introductory biology texts at the university level in Canada and the United States.

6. Paul (1995) makes an interesting observation in her paper on textbook presentations of the genetics of intelligence. Although she did survey 28 of the 31 texts available to her at the time, after completing the research, she concluded: "It does not appear that the findings would have differed in any significant respect had the study been limited to what are *apparently* [author's emphasis] the most popular texts" (319).
7. I discovered that there was no database that indicates text usage or "market share" of each text in the OAC biology "market." For example, the Faculty of Education at the University of Toronto had kept records until 1997. Even these older records were no longer accessible, however.
8. Altbach has indicated that the system of textbook production is dominated by a relatively small number of countries and their "major academic institutions, research laboratories, and publishers" (Altbach, 1991, 244). According to Altbach, "Canada, for example, has long been concerned that many of the books used in its universities have little, if any, 'Canadian content' and present material that is not relevant to Canadian readers" (254).
9. I did not have sufficient resources to conduct a detailed survey of high schools in Ontario to ascertain which books were being used in each case. I spoke informally, however, with administrators, science consultants and teachers who attended the Science Teachers Association of Ontario annual meetings in the Toronto area in 1999 and 2000 and the Science Coordinators Association Ontario meeting in Barrie in the spring of 2000, and through telephone conversations with representatives from publishing houses.
10. Galbraith also edited *Biology Directions* (1993) on the interim-approved Circular 14 list. However, this text does not include a chapter on animal behavior and sociobiology.
11. The Raven and Johnson text that was on the interim approval list was created in two parts as a special edition for OAC courses. I was unable to find Part II of this special edition and instead used a newer, one-volume edition supplied by the new publisher. I felt that this was acceptable because the newer text seemed to be as popular and was the only edition available to schools at this time.
12. A representative of the William C. Brown company, the publishers of the Raven and Johnson text, estimated that it has been widely used in southern Ontario OAC biology courses and also widely used in a modified version in at least 30 percent of U.S. introductory biology courses. The Niagara School District, for example, gave teachers a choice only between the Galbraith text and the Raven and Johnson text.
13. It is interesting to note that Miller, one of the authors of this text, has gone on to write two books (2008; 1999) on controversies surrounding evolutionary theory.
14. Also, the texts now being produced for the new, lower-level biology 12 programs in Ontario are entirely Canadian products and much less extensive than either the U.S. or Canadian texts that were used in the grade 13 programs. They do not deal much with issues of biology and behavior.

· 5 ·

THE RAVEN AND JOHNSON TEXTBOOKS

In this chapter I do a textual analysis of two related textbooks. The first is *Understanding Biology* (3rd edition), written by Peter H. Raven and George B. Johnson and published by William C. Brown Publishers (1995). It is in most ways the most pro-Wilsonian text of all that I examine. The second is the eighth edition, called *Biology*, that is based on the work of Raven and Johnson.

Part 1: The Third Edition

Presentations of Animal Behavior and Sociobiology in the Third Edition

The third edition of *Understanding Biology* is a large, multicolored edition with inserts, graphics, and a CD-ROM. Chapter 45, titled "Animal Behavior," is the last chapter in the book. This seems to be a common placement for this topic. It contains sixteen pages of text and pictures, two pages of summary, and review material. The chapter opens with a full-page, full-color photo of a group of African meercats standing at attention. It sets the tone for the chapter and echoes the overall emphasis of the text. The caption reads:

> On the lookout. These African meercats are acting as sentries, perched on top of high termite mounds where they can see approaching danger. It is an interesting biological question why some individuals of a group will expose themselves to danger like this to benefit others. Does evolution favor such behavior? (853)

This insert introduces one of the most important "problems" that sociobiology sets forth to solve—the problem of altruistic behavior. The use of this image at the beginning of this chapter clearly indicates what the authors consider to be important theory and issues in discussion of animal behavior. This image foreshadows the discussion of altruism that will follow in the chapter's discussions about sociobiology. This concept is one of the most studied behaviors in sociobiology, and research on the meercat is supposed to confirm the claim that natural selection does favor the evolution of altruistic behavior in conjunction with the operation of kin selection.[1]

Chapter 45 defines behavior, in the first paragraph of the chapter, as "the way an organism responds to a stimulus in its environment" (854). In the third paragraph of the chapter, the authors go on to say that

> animal behavior can be explained in two ways: The first involves *how* the animal's senses, nerve networks, or internal state provide a physiological basis for the behavior. Analysis of the *proximate cause*, or mechanism, of behavior, involves measuring hormone levels or recording the firing patterns of neurons. The second explanation for animal behavior involves asking *why* the behavior evolved; that is, what is its adaptive value? Study of this *ultimate*, or evolutionary, *cause* of a behavior involves measuring how the behavior influences the animal's survival or reproductive success. Thus a male dog's frenetic activity when he wants to mate can be explained by hormones, internal messengers released in the spring that cause him to seek out females; this is the proximate cause for his behavior. Evolutionarily speaking, however, the dog shows this behavior to pass on his genes. In effect, genes guide the dog to make more genes; this is the ultimate cause. (854)

This is a clear and concise statement of the core reasoning in sociobiology. This forcing of the genetic-adaptationist linkage is characteristic of sociobiology and other forms of reductionist determinism. There are behavioral states in animals generated by biochemical processes that are linked to genes, and these genes are "using" the animal and its behavior to "make more genes" in conjunction with adaptations to the dictates of selection pressures. The authors go on to indirectly delineate a dominant or hegemonic position for sociobiology and other evolutionary studies of behavior in the next paragraph.

> This chapter considers the mechanisms by which animals respond to their environment, as well as the adaptiveness of their behavior. Scientists have taken different

approaches to the study of behavior, some focusing on instinct, some emphasizing learning, still others combining the elements of both approaches. The picture that emerges is one of behavioral biology as a diverse science that draws strongly from allied disciplines, such as neurobiology, physiology, psychology, and ecology. *Today, the overarching theme of the study of behavior is evolution* [author's emphasis]. The chapter discussion of human behavior will demonstrate how such an evolutionary perspective can be controversial when applied to the social behavior of humans. (854)

These passages reflect many of the problems that critics have highlighted since the inception of sociobiology and the related field of behavioral genetics. One in particular is the problem of the movement from proximate causes that involve gene action, protein synthesis, and measurable biological structures to "ultimate causes" involving the intentional action of genes as agents of evolutionary change. I covered these issues and critics' responses in detail in Part I of this book. As I indicated in Chapter 1, what this reductionism accomplishes is the disappearance of the entity as a complex organism. A dog becomes the sum total of its genes in an abstracted way. When this happens, all of the complexity that involves the movement from genotype to phenotype can be lost. And the existential and social dimensions with which the organism contends also will be lost. In addition, in this last passage we see the ultimate movement from nonhuman to human examples. This is reminiscent of Wilson's (1975) use of the last chapter in *Sociobiology: The New Synthesis* to introduce human sociobiology. Once the reduction has been established in animals, it is then applied to humans.

The hegemonic position for evolutionary and genetic models of animal behavior is carried through in the first section, titled "Approaches to the Study of Behavior." In this first section of the chapter, the text offers a contradictory message about the roles and importance of biology and environment. First, the authors outline the nature/nurture debate in only one paragraph. The authors take the reasonable position that

although in the past, the nature/nurture controversy has been an "either/or" argument, many studies have shown that both instinct and learning play significant roles, often interacting to produce the final behavior product. The scientific study of instinct and learning, as well as their relationship, has led to the growth of several disciplines, such as ethology, behavioral genetics, behavioral neuroscience, and psychology. (854)

What follows immediately from the quote given above is a two-paragraph presentation of the general principles of ethology.

In the chapter, ethology is a subsection of "Approaches to the Study of Behavior." The ethology subsection is further subdivided into "The Genetic

Basis of Behavior" and "The Neural Basis of Behavior." This last subsection delineates the field of neuroethology in half a page. There is no presentation of what critics would call environmental variables in the ethology section. The next subsection of "Approaches to the Study of Behavior" is titled "Psychology and the Study of Animal Behavior." This subsection discusses associative and non-associative types of learning in one page. They then spend the remaining one and a half pages on "The Genetic Aspects of Learning," "The Physiology of Behavior," and "Behavioral Rhythms."

In the introduction and the "Approaches to the Study of Behavior" section it is not so much the specific presentations around genes, biology, and behavior that I found problematic but more the overall impression with which the reader is left. I will try to detail this impression in the following pages. To their credit, the authors begin the "Genetic Aspects of Learning" subsection by reminding the reader "that behavior has both a genetic and learned component" (858). In the next subsection, titled "Genetic Aspects of Learning," they conclude that genotypes set a range of reaction to specific stimuli. And so, "Animals are innately programmed to learn some things more readily than others" (858). Furthermore, Marler's work on bird song

> suggests that birds have a *genetic template*, or instinctive program, to guide learning the appropriate song. Song acquisition is based on learning, but only the song of the *correct* species can be learned. The genetic template for learning is *selective*. (858)

This "Genetic Aspects of Learning" subsection concludes with a colored, highlighted insert

> Single genes are not responsible for complex behaviors, and complex behavior is not entirely governed by learning. Genes influence what can be learned. The way that genes and experience interact is adaptive. In different species, instinct and learning vary in importance, but in many species, genes seem to set the limits on the extent to which behavior can be modified. (858)

To their credit, the authors have not misrepresented the work they describe, and I do not think that any of these statements is an over-exaggeration of the kind of behaviors that can have genetic and biological influences. I find, however, that the presentation of the causes of behaviors in animals in the introduction and first section of this chapter overwhelmingly convey to the reader the sense that genetics, physiology, and selection are the most important factors. This impression is created as much by what is not said as by what is said. This textbook presentation highlights the issue I discussed earlier concerning the marginalization of the various levels of an organism's reality.

Despite the authors' continued qualifications concerning the influences of biological and environmental factors, six and a half pages of the combined introduction to this chapter and the "Approaches to the Study of Behavior" section are devoted to biological influences. At the same time, only one and a half pages are devoted to learning or environmental influences. This appears to constitute the process of "mentioning" that I discussed in Chapters 2 and 4. "Mentioning" serves two functions. It creates the appearance of evenhandedness, and it also serves to muddy issues and create confusion and ambiguity in the minds of readers. Ultimately, if you actually leave little room for nongenetic aspects of learning, readers confront a confusing portrayal in which they are told that both biology and learning matter, while in the main only biological and genetic factors are given credence. There is a kind of doublespeak, or at least possible confusion, in this section about how to conceptualize the many levels of influences that can be considered. This confusion seems to me to have the effect of rendering the more experiential components invisible or at least relatively unimportant.

This is quite typical of Wilson's sociobiological discourse about the impact of nature and nurture in which he appears to pay lip service to the dual roles of nature and nurture in the determinations of behavior. It is almost always followed by a complete disregard for environmental influences in the ensuing discussion of the genetic and biological influences that are seen ultimately to control behavior. It usually is also connected with an almost complete disregard for the complexities of an organism's experience and for the complexity of traits such as intelligence and personality.

Another rhetorical strategy present in both sociobiological discourse and this text is unqualified movement from nonhuman to human examples. This is evident in the next major section, titled "Animal Communication." This section contains about one and a half pages of text and one page of illustrations. This section has subsections titled "Courtship," "Communication in Social Groups," and "Human Language." One-quarter page of text is devoted to the "Courtship" and "Communication in Social Groups" subsections. About one half-page of the text portion is devoted to an insert (Sidelight 45.1), called "The Dance Language of the Honeybee" (862). This insert details the history of research conducted to understand how the honeybee "dance" communicates to other members of a hive where to find food. The authors do not make any attempt here to link this behavior to either genetics or experience. Again, this silence is as significant as what is said in the chapter and I will discuss it again below.

One half-page of text is dedicated to the subsection on "Human Language." The first two subsections discuss social releasers and stimulus/response chains for things such as species-specific courtship signals, alarm pheromones in ants, and honeybee dances. The subsection on human language is the first mention of human behavior in the chapter since the sentence from the first page of the chapter that I quoted above. It is inserted without any qualifying remarks as to how the study of human behavior may differ from that of ants, bees, fish, and birds. In this subsection, the authors claim that

> although human languages appear on the surface to be very different, in fact, they share many basic structural similarities. Researchers believe that these similarities reflect how our brains handle abstract information, a genetically determined characteristic all humans share. (860)

There has been an unqualified movement from claims and extrapolations about the behaviors of insects, birds, fish, or lower mammals to claims and extrapolations about humans. Like Wilson's sociobiological discourse, there is no serious grading or distinguishing of organisms and their biology and environments.

As I discussed in Part I of this book, one of the mechanisms that sociobiological discourses and general determinist formulations use to create a conceptual framework for themselves is to transform any manner of behavior into a characteristic or trait without defining any ground rules for this process. By calling behavior "adaptation," in the definition above, behavioral ecology grants the right for any behavior to be treated as a characteristic or trait and then fit into a simple Mendelian and adaptationist paradigm. This process can be seen in the next section of the chapter, called "Ecology and Behavior," which comprises three pages of text, photos, and inserts. In the first paragraph of this section, the authors introduce the reader to the concept of *behavioral ecology*. It is defined as the study of "the ways in which behavior serves as adaptation and allows an animal to increase or even maximize its reproductive success" (862). The text then goes on to discuss orientation and migration in birds and insects, foraging behavior, territoriality, reproductive strategies, mate choice, and sexual selection as some of the behaviors studied by behavioral ecologists.

The last major section of the chapter is called "The Evolution of Animal Societies." It contains five pages of text, graphics, and inserts. The first subsection of this major section is called "Sociobiology: The Biological Basis of Social

Behavior." This is the first direct mention of sociobiology in the chapter. The text presents Wilson, Alexander, and Trivers as the founders of "what has become a major movement within biology" (866). It defines this movement in the following way:

> A biological view of social behavior being the result of evolution would predict that the behaviors characteristic of particular animals are, by and large, suited to their mode of living—that is, in the sense of Darwin, are adaptive. Natural selection would be expected to favor those gene combinations that allow animals to adapt more completely to particular habitats. The study of the biological basis of social behavior of animal societies is called *sociobiology*. (866)

The text then goes on over the next four pages to detail some of the behaviors that have been studied by sociobiologists and some of their key principles. It covers altruism, reciprocity, inclusive fitness, and kin selection. Inclusive fitness is defined as "the sum of the genes propagated by personal reproduction and by the effect of help with relatives' reproduction" (867). According to the text,

> the theory of kin selection proposed by W. D. Hamilton predicts that altruism is likely to be directed toward close relatives. The closer the degree of relatedness, the greater the potential genetic payoff in inclusive fitness. (867)

The text then looks at behaviors in both insect societies and in vertebrate societies. In the latter, it looks at cooperative breeding, alarm calling, and mating systems, wherein it concludes in a colored, highlighted insert

> Social behavior in vertebrates is often characterized by kin-selected altruism. Altruistic behavior is involved in cooperative breeding in birds and alarm calling in mammals. The organization of a vertebrate society represents an adaptive response to ecological conditions. (870)

Then, in a manner reminiscent of Wilson's text, the last subsection in the chapter is titled "Human Sociobiology." The authors immediately qualify this discussion by noting: "[A]re humans just another social animal whose behavior can be explained and understood in Darwinian concepts?" (870) They go on to catalog some of the range of complex human behavior and note the following

> If an ethologist took an inventory of human behavior, he or she would record kin-selected altruism and reciprocity, other elaborate social contracts, extensive parental care, conflicts between parents and offspring, violence, and warfare. A variety of mat-

> ing systems, such as monogamy, polygamy, and polyandry, would be described, along with a number of sexual behaviors, such as adultery and homosexuality. Behaviors such as adoption that appear to defy evolutionary explanation would also be a part of the ethologist's catalogue.…Are these behaviors rooted in human biology? (870)

The authors then go on to attempt an answer by bringing in the concepts of both biological evolution and cultural evolution as components that create human societies. Inclusion of the latter indicates that the authors are versed in some of the modifications that have been made to Wilson's early formulations. They then ask the question: "How can biological components of human behavior be identified?" They answer it by suggesting that one "study behaviors that are cross-cultural [which]…[i]n spite of cultural variation" are traits that "characterize all human societies" (870). The example they give of just such a "trait" is the incest taboo. And they conclude that "Natural selection may have acted to create a cultural norm to avoid a serious biological problem. Genes responsible for guiding this behavior might have been fixed in human populations because of their adaptive effects" (870).

This chapter wholeheartedly embraces sociobiology and behavioral ecology as further representation of the power of evolutionary theory and modern genetic theory. Fourteen and one-half of the sixteen pages in this chapter are devoted directly to the evidence for the biological, genetic, and evolutionary bases for behavior. Half of the chapter (eight pages) is devoted to direct sociobiological and related formulations. What is not said is equally important, however. The authors pay lip service to the duality of biological and cultural evolution without providing the reader with any way of conceptualizing this duality. In the penultimate paragraph of the chapter, the authors do say that "A significant number of biologists and social scientists vigorously resist any attempts to explain human behavior in evolutionary terms" (870). None of the questions and criticisms that have been raised about the validity of the sociobiological exercise has been presented, however. The chapter concludes with the following passage:

> Moreover, if human behavior is considered to be the product of evolution—influenced by genes and at least in part "hard-wired"—does this not suggest that unpleasant aspects of human behavior such as aggression and violence, cannot easily be modified? Such a view could affect how we perceive the prospects for positive social change.
>
> Darwinian theory can provide an overarching evolutionary *perspective* on human nature.…But human nature is affected by *both* innate and learned components, and because many human activities…are strongly influenced by culture and are not easy to study as adaptation, Darwinian theory is unlikely to offer any *resolution* of the fine details of human nature. (871)

What then follows is an "Evolutionary Viewpoint" insert in which the authors conclude that

> the evidence now seems overwhelming that genes dictate or influence many if not most aspects of human behavior and personality. Heritability[2] studies comparing identical and nonidentical twins raised apart and raised together clearly indicate that aggressiveness, intelligence, and many other key attributes are more influenced by what genes we inherit than by how we are raised. However, this does not mean that genes determine everything and that how children are raised and educated does not matter. Mendelian segregation only ensures that every child will be different, a unique challenge and opportunity. (871)

What is incredible about this "viewpoint" is that it bears no direct connection to the material on human sociobiology covered in the chapter. There are no data presented in the body of Chapter 45 that in any way support the above statement. The authors present no evidence in support of human sociobiology or scientific debate about its accuracy. Instead, they give a page of proclamations about its logical consistency and its inevitable correctness. In the end, my overall impression is that this chapter leaves the reader with an overarching sense that human behavior must be directly under genetic control, while providing no sense of alternative positions and no real sense of how to evaluate these claims.

Presentations of Genetics in the Third Edition

Unit 4 of this text is devoted to genetics. There are no overt discussions of behaviors as traits and no overt discussions of genetics and behavior in general. The text makes very little attempt to present or discuss the nature/nurture debate directly in its presentations on genetics. It does, however, take a very clear position on this debate in indirect ways. In fact, the text does a very good job of presenting the complexities of genetic interactions. At the same time, though, it does a poor job of qualifying or framing considerations of genetic influences within a wider theoretical framework that includes nongenetic influences. In ways it reinforces the process of reification that operates in Wilsonian human sociobiology. The chapters on genetics themselves support the strong linkage of genes with human behaviors. The authors pay lip service to the issue of nongenetic influences on behavior, while simply omitting any real discussion of environmental considerations and reinforcing the legitimacy of genetic determination in the human realm. All of this is done in conjunction with the use of a number of rhetorical devices that also reinforce a strong genetic presentation of the influences on human behavior. These devices include the process of "mentioning," unqualified sliding from nonhuman to human examples, and the use of pictures of human examples that appear in some sense to be "outside" the main body of text. All of this is done with no real discussion of environmental influences, no adequate definition of "traits," and the implicit promotion of a process of reifying behaviors as traits. I detail these below.

This unit begins with Chapter 9, titled "How Cells Reproduce." This chapter covers issues of cellular reproduction, both mitotic and meiotic. It also covers various forms of reproduction and gives details on the human sexual life cycle. It also provides a good deal of detail about reproduction at the chromosomal level and includes much detail on the cellular mechanics of reproduction. Chapter 10 is titled "Mendelian Genetics" and begins with a full-page color photo of twin boys. The insert caption reads:

> "Some of us look more alike than others." These identical twins have all their genes in common, and differ only in how they have developed. In general, people differ from one another in a significant fraction of all their genes, and these so-called "heritable differences" are responsible for *much of the differences among us in appearance and behavior* [author's emphasis]. (175)

This is a very clear statement that links basic Mendelian principles, genes, behaviors, and humans. It is also an example of the "sliding" of human examples into a more general discussion of genetic concepts and also an example of the use of photos and inserts to highlight issues around the connections between genetics and behavior. It begins the chapter and frames the issues for the reader in a very direct way. It also does this outside the body of the text in a photo and caption. This format occurs more than once in the chapters on genetics. This use of human examples, human images, and human twins sends a subtle message that reinforces the importance of genetics in the determining of human behaviors. At the same time it creates a link between human "common sense" notions about the influences of heredity and Mendelian genetics. As I discussed in Chapter 2, this bonding of common sense and science claims is an important keystone to the development of legitimacy and popularity for genetic explanations of human behavior.

All of these currents come together around the presentations of traits. As is the case with most texts, this term is left undefined, unqualified, and vague. There are no listings in the index for either "characteristic" or "trait" and no listings in the glossary of the text for either concept. Each chapter begins with a "For Review" insert that instructs the reader: "Here are some important terms and concepts that have been discussed in previous chapters and that you will encounter again in this chapter. Review them before proceeding if necessary" (308). "Characteristic" and "trait" are never highlighted in this way. In this text they appear as concepts so thoroughly accepted as to be invisible or, at this point, completely naturalized. Without any clear definition or attempt at qualification of what can count as traits, the text leaves open to the reader the possibilities of conceiving any complex human behavior as a trait and therefore subject to basic Mendelian and Darwinian controls.

The rhetorical devices of "mentioning" and the specific use of photos and inserts to emphasize specific points and to "slide" from nonhuman to human examples are all evident in Chapter 10. For example, Figure 10.2, titled "Inherited Traits," is a photo of a small child with red hair. The caption reads: "Red-haired parents often produce children with red hair" (176). Likewise, Figure 10.10, called "A Common Recessive Trait in Humans," features a picture of a blue-eyed man with a caption that reads: "Blue eyes are considered a recessive trait in humans, although many genes influence the exact shade of blue" (182). As with the photo of the twins, these photos all serve to legitimize the linking of traits with genes in humans.

Then, in Chapter 11, titled "Human Genetics," the authors stay away from specific discussions of genes and behavior, but the topic is again indirectly addressed. This chapter covers the details of human genetics and presents data on many genetic abnormalities and genetically based diseases such as sickle-cell anemia, Huntington's disease, phenylketonuria, hemophilia, cystic fibrosis, and Tay-Sachs disease. The text also discusses pedigree analysis and briefly, at the end of the chapter, introduces the processes of genetic counseling and genetic therapy. In the midst of these presentations is a highlighted insert called "Sidelight 11.1: The Human Genome Project." In this sidelight, the authors ask the question: "What if the scientists working on the Human Genome Project discover a gene involved in determining intelligence? Will everyone be tested for this gene? Will those who do not have the correct gene be eliminated?" (199). In a fashion reminiscent of Wilsonian discourse, the use of a "what if" form introduces—without evidence—the implicit supposition that intelligence can be a trait directly influenced at the genetic level. In this case, there is no discussion of the possible mechanism of such inheritance; there is only a discussion of the ethical issues involved in the identification of such genes—as if this is already scientific fact.

In addition, the last two and a-half pages of Chapter 10 deal with the topic called "From Genotype to Phenotype: How Genes Are Expressed" (189). As with discussions of traits, the presentations of genotype/phenotype interaction are important to the framing of the general genes and behavior debate. As I indicated above, this text often "speaks without speaking" through its almost complete omission of any real discussion of non-genetic factors in phenotype emergence. For example, in Chapter 10, the authors go through the concepts of multiple alleles, epistasis, continuous variation, pleiotropy, and incomplete dominance in very thorough and informative ways. Figure 10.18, in Chapter 10, is called "Height as a Continuously Varying Trait." This figure contains both an old photo of students arranged by heights posed so as to mimic a frequency distribution and a bar graph illustrating the distribution of individuals and heights. The caption reminds the reader that

> because many genes contribute to height and tend to segregate independently of one another, many combinations are possible. (b) The cumulative contribution of different combinations of alleles to height forms a spectrum of possible heights—a random distribution in which the extremes are much rarer than the intermediate values. (191)

In this example, the authors demonstrate the complexities *internal* to the level of genetic interactions that can affect height. They do not, however, discuss environmental variables and their influences on height.

The chapters on genetics in this text offer the reader little serious or thorough discussion of environmental or nongenetic influences on organisms—both human and nonhuman—and on their behaviors. Rather, they tend to mention environmental influences as in Figure 10.1. This is a photo of ten babies lying on mats. The heading reads: "No two individuals look alike." Its caption reads: "As a result of both heredity and environmental influences, these babies will grow to adulthood as separate individuals" (176).

Likewise, the associated section on "environmental effects" in the discussion of genotype/phenotype interactions in this chapter is exactly four lines long and tells the reader: "The degree to which many alleles are expressed depends on the environment. Some alleles encode an enzyme whose activity is more sensitive to conditions such as heat or light than are other alleles (figure 10.20)" (191). In Chapter 11, the authors also stay away from discussions of environmental influences. The only reference to environmental factors occurs in the second paragraph: "The ways in which genes interact with the environment to produce individuals with specific characteristics are the subject of continuing study." In Chapter 12, titled "DNA: The Genetic Material," we see genetics at the molecular level. It provides rich detail on the structure and function of nucleic acids. It is very well done and contains inserts that discuss key experiments in the history of molecular genetics. It presents data on many kinds of living organisms from viruses to humans. It does not focus on issues of genetics and behavior except in Figure 12.1, titled "Heredity Shapes All of Us." This figure is a photo of a boy with his grandfather. The text accompanying this photo reads:

> *Some* [author's emphasis] of what this boy will be as an adult will be influenced by what he learns from his grandfather, but *much* [author's emphasis] will reflect the genes he has inherited from his grandfather. (218)

The authors are not saying anything "incorrect" in these inserts. Again, we see two recurrent patterns in this text. With one pattern the weight on pronouncements about genes and behavior is always skewed towards the genetic influences. With the second pattern, there is a brief qualification of genetic influences and appeals to consideration of environmental influences.

This kind of "mentioning" of environmental influences indicates the superficial way in which nongenetic factors are treated in this text. In the absence of any qualifying comments about genes and environments, these inserts can have the effect of moving the reader toward a strong genetic determinism—especially when counterposed with the few unsystematic and undeveloped comments about nongenetic influences on human behavior. The net result of the use of these photos and inserts is that points are made without direct reference to the issues in the body of the text. These photos and inserts support a key component of the reification and reduction process. They accomplish the slide from nonhuman to human examples and, at the same time, they mention environmental influences. And ultimately, they link a common sense experience of relatively simple biological traits, such as hair color, with the implication that other more complex "traits" in humans can also have genetic origins. The general tendency of the authors in this unit on genetics is to do a fine and detailed job of presenting the complexities involved in understanding genic interactions, especially at the molecular level. Yet there is very little analysis done in this text that would allow teachers and students to develop a coherent and complex position on the interactions of genes, environment, cognition, and culture in the production of human behaviors.

Presentations of Evolutionary Theory in the Third Edition

This text gives prominence and a hegemonic place to evolutionary theory.[3] To demonstrate this, I would point the reader to the first chapter. Chapter 1 is titled "The Science of Biology." It begins with a full-page, color photo of a very old bristlecone pine tree. In addition to the photograph, there is a text insert that reads:

> This bristlecone pine has lived for hundreds of years in a harsh climate—the successful result of eons of evolution. Evolution is the core of the science of biology. (3)

The next six pages are then devoted to a section titled "History of a Biological Theory: Darwin's Theory of Evolution." The prominent place occupied by evolution within the theoretical space of the first chapter signals what will follow. I do not see this as necessarily positive or negative. Given the kinds of challenges that evolutionary theory has had to defend against in some parts of the U.S., it is entirely possible that the authors give evolution such prominence as a defense against creationism and other religious-based challengers.[4] This idea is supported by an insert on the summary page of the chapter called "Evolutionary Viewpoint." This insert says:

> The core explanatory principle of the science of biology is evolution by natural selection, a theory first advanced by Charles Darwin over one hundred years ago. Diverse lines of evidence convince biologists of the general validity of this theory, although lively discussions continue about the details. There is essentially no support among biologists for so-called "scientific creationism," which holds that the biblical account of the origin of the earth is literally true, that the earth is much younger than most scientists believe, and that all the species of organisms were individually created just as they are today. (19)

In the author's own words,

> those familiar with the first two editions of *Understanding Biology* know that evolution has been the grounding feature of this text. Evolution remains the guiding perspective of this third edition, and our organization of the form and function chapters, along with the new *Evolutionary Viewpoints* at the end of each chapter, reflect this perspective. (xxiv)

These evolutionary viewpoints serve to reaffirm the dominant position that evolutionary theory has as an interpretive framework for all areas of biology. So, for example, the highlighted box at the end of Chapter 9, titled "How Cells

Reproduce" discusses the role of synapsis that together with crossing over "has been a key factor in the evolution of eukaryotes because it is a powerful mechanism for reshuffling genes" (173).

Unit 5 of the text is called "Evolution." It is composed of Chapters 15 through 19. Chapter 15 is called "The Evidence for Evolution." As I indicated above, the authors rightly see one of the aims of the text as that of combating challenges to the hegemony of evolutionary theory that come from religious positions. They make this clear in this chapter that opens with the following paragraphs:

> Almost everybody has heard of Darwin's theory of evolution—not because of widespread interest in biology but because many people regard this theory as a challenge to their religious beliefs. . . .
>
> Within recent years, controversial attempts have been made to require the teaching of the set of religious dogmas known as scientific creationism alongside evolution theory in science classes. The theory of evolution has faced similar highly publicized challenges since the time of Darwin, and others are likely in the future. For this reason, this chapter addresses the issue squarely. Just what is the evidence for evolution? (286)

The authors mount a very thorough and organized presentation of the evidence that supports the theory of evolution. They discuss population genetics, effects on allele frequencies, forms of selection, microevolution, and the evidence for macroevolution in great and clear detail.

They present a very clear picture of the forces that affect the frequency of alleles in a population. They discuss mutation, migration, genetic drift, nonrandom mating, and natural and artificial selection. They do this in a way that maintains the complexity of the interactions of genotype and phenotype in combination with selection pressures. For example, in the passage on selection, they explain that

> Darwin argued that the more successful reproduction of particular genotypes, which is how he defined selection, is the primary force that shapes the pattern of life on earth. But the selection of these genotypes is indirect: Selection acts directly on the phenotype, which is determined by the interaction of the genotype and the environment, and the linkage between particular alleles and particular characteristics of the phenotype is less direct for some features than for others.
>
> Although selection is perhaps the most powerful of the five principal agents of genetic change, there are limits to what it can accomplish. These limits arise because alternative alleles may interact in different ways with other genes. These interactions tend to set limits on how much a phenotype can be altered. . . .

> A second factor limits what selection can accomplish: Selection acts only on phenotypes. Only those characteristics expressed in an organism's phenotype can affect the organism's ability to reproduce progeny. Selection does not operate efficiently on rare, recessive alleles simply because they do not often come together as homozygotes, and there is no way of selecting them unless they do come together. (290–291)

What is interesting about this presentation is that the authors do not appear to be presenting an ultra-adaptationist position. Rather, they seem to take pains to create an appreciation for the complexities involved in the process. In Part I of this book, I discussed the ways in which the Wilsonian discourse of human sociobiology relied on an ultra-adaptationist version of evolutionary theory to help bolster its legitimacy and credibility.

Chapter 16 is titled "How Species Form." This chapter details the evidence for how species form. It outlines concepts such as fitness and discusses the factors that are barriers to species hybridization. Chapter 17, titled "Evolution of Life on Earth," deals with the fossil record that supports the theory of evolution. Again, it presents a thorough delineation of all that we know about the ages of the earth, continental drift, and carbon-14 dating and the fossil record. This would normally be seen as more relevant to a geography text, but undoubtedly the authors include it here to convince readers that the earth is indeed many billions of years old and that the fossil record supports the theory of evolution. Chapter 18 is called "The Story of Vertebrate Evolution" and details the evidence for this evolution. This chapter details the archeological evidence for primate evolution, from *Homo habilis* to *Homo erectus* to *Homo sapiens*. The chapter goes into detail discussing the studies conducted using mitochondrial DNA and discusses Neanderthals and Cro-Magnons. This chapter ends with an interesting and somewhat incongruous (for this text) passage:

> While not the only animal capable of conceptual thought, we have refined and extended this ability until it has become the hallmark of our species. We use symbolic language, and with words, can shape concepts out of experience. This has allowed the accumulation of experience that can be transmitted from one generation to another. Thus, we have what no other animal has ever had: cultural evolution. Through culture, we have found ways to change and mold our environment to our needs, rather than evolving in response to environmental demands. We control our biological future in a way never before possible—an exciting potential and a frightening responsibility. (376)

I say this is an incongruous passage because it seems to allow for the impact of human culture on human behavior, human society, and human evolution.

This is not an idea that was developed in any detail in discussions of animal behavior and sociobiology, however. As with the case of numerous issues in many of these texts, this issue is only "mentioned" and the net impact of the comments, I feel, is likely to be insignificant to a reader.

Presentations of the Nature of Science (Knowledge) in the Third Edition

The only place where the text deals directly with the issues of the nature of scientific knowledge is in Chapter 1. There are three and a half pages on what the text calls "The Nature of Science." This section deals with showing the reader what the method of science investigation is and demonstrating how hypotheses evolve into theories through the accumulation of evidence gathered through experimentation. This presentation is neopositivist in the sense that it does not discuss any complex epistemological issues around the process of knowledge creation and does not try to place the development of theory in any possibly social context. In the point summary page of Chapter 1, however, the authors do say that, "because even a theory is accepted only provisionally, there are no sure truths in science, no propositions that are not subject to change" (19). In this way, the text appears to tend to support a neopositivist position, but the simple fact that they have a section titled "The Nature of Science" shows that they are aware of the issue.

This text does not deal with any constructionist discussions of the nature of science directly. It maintains the boundaries of debate firmly within the neopositivist orbit. The authors do take on the claims of scientific creationism directly in a section at the end of Chapter 16. They tell the reader that science and religion can coexist, and that

> science provides a coherent means of organizing observations, and of making predictions about how the world is going to behave. It is not a substitute for religion, which addresses a different arena of human concerns: questions of ethics and ultimate causes. Religion and science do not preclude one another but are regarded by many as complementary ways of viewing the world. (302)

And

> Scientific creationism should not be labeled science for three reasons: (1) It is not supported by any scientific observations; (2) it does not infer its principles from observation, as does all science; and (3) its assumptions lead to no testable and falsifiable hypotheses. (303)

This is an interesting set of passages, because in dealing with the issue of creationism, the authors also deal with questions about the nature of science and what makes it different from other systems of knowledge, such as religion. They present only a neopositivist account of the nature of science, however. This is, in some ways, understandable, as their aim is to differentiate religious belief from a materialist account of reality. It requires a great deal of work to present and demonstrate how a constructionist framework can be seen as a materialist account of reality.

The only other place where one can infer a treatment of the nature of science is in the text's treatment of Mendel. Raven and Johnson's presentation of Mendel is extremely canonical. Chapter 10 presents Mendel's research on pea plants. It is a clear and thorough presentation of the reasons why Mendel chose the experiments he did, and the reasons for his conclusions. It is a very thorough exposition of Mendelian principles. It is also, however, cast as an instructional lesson for students as an appropriate example of the proper use of the scientific method. For example, one insert says:

> Gregor Mendel chose peas for his classical genetics experiments because the results of crosses in peas had been studied earlier; different varieties are variable in their features; the plants are relatively small, so many can be grown in a limited area, and the generation time is short; and true breeding strains are produced by self fertilization. (178)

This is a very different presentation from that presented by Galbraith (1993), which attempts to contextualize Mendel's work and which I discuss in Chapter 7. As I indicated in Chapter 4, there is some controversy around the accuracy of the data reported in Mendel's experiments and his status as the "founder of modern genetics." The texts have an opportunity in the presentation of canonical figures to introduce the students to the idea that science method and claims are more socially situated than most neopositivist accounts maintain. This text does not take advantage of that opportunity, however.

Summary of the Third Edition

All in all, the third edition text supports common sense linkages that favor support for a strong program of genetic influences in general, and, by omission, a strong program for genetic influence on human behaviors that can be conceived as traits. This is especially true in the sections that deal directly with animal and human behavior. It is also true of some of the material covered in the sections on genetics in the text, however. Discussions slide from nonhuman example to human, and there is little discussion of non-biological factors in behavior and in the emergence of a given phenotype. Likewise, there is very little discussion of the nature of scientific knowledge, and Mendel and Darwin tend to be treated as canonical figures extracted from their social and historical contexts. In these ways it is the most similar in general argument, structure, and emphasis to Wilsonian sociobiological discourse.

Part 2: The Eighth Edition

The eighth edition is titled *Biology* and it is in the hands of new authors. The first inside page informs us that the new authors are Losos, Mason, and Singer, and that the text is based on the work of Raven and Johnson. Losos and Singer served as co-authors for the seventh edition; all three became the principal authors in 2008 in the eighth edition (Losos et al., 2008; v[5]).

Presentations of Animal Behavior and Sociobiology in the Eighth Edition

It is interesting to note that in the fifth edition of Raven and Johnson's *Biology*, published in 1999, the chapter on animal behavior is split into two chapters. The first, Chapter 59, is called "Animal Behavior" and deals with, in effect, pre-sociobiological theories about animal behavior. Chapter 60, titled "Behavioral Ecology," is devoted to analyzing "how evolution has shaped, and is shaping, the behavior of animal species in natural populations" (Raven and Johnson, 1999, xvix). This is the chapter in which sociobiology is discussed. In a clear indication of how these authors feel, the more recent textbook ends with the following quote that is highlighted and in boldface.

> Sociobiology offers general explanations of human behavior that are controversial, but are becoming more generally accepted than in the past. (1226)

However, by the eighth edition the position has changed. The chapters on animal behavior in the eighth edition are very illustrative of these overall changes. Behavior is given a whole unit in the text, consisting of six chapters. The most relevant chapter for this work is the first chapter in the unit, Chapter 54, titled "Behavioral Biology." The usual material is presented, and areas such as innate behaviors, proximate and evolutionary causes of behavior, behavioral genetics, instinct and learning, etc., are all covered well and thoroughly. It is significant that in the section of the chapter titled "The Development of Behavior" (1122–1123), the authors present two subsections. The first is called "Parent-Offspring Interactions Influence Cognition and Behavior" (1122–1123). The second is titled "Instinct and Learning May Interact as Behavior Develops" (1123). In both of these subsections, the authors present complex formulations for the ways in which animal behaviors emerge through instinctual drives and genetic predispositions that are mediated through social relationships and through learning. This treatment is more than "mentioning" and it includes pages of discussions.

What is most striking about this chapter on animal behavior in the eighth edition of this text (based on Raven and Johnson's earlier texts) is that there is no use of inserts to introduce human examples into the discussions. In fact, human behavior is only mentioned in one very brief section comprising only two paragraphs. This subsection is called "Human Twins Studies Reveal Similarities Independent of Environment" (1118), in the chapter section called "Behavioral Genetics" (1117–1120). In this small subsection, the authors tell the reader: "The role of genetics can also be seen in humans by comparing the behavior of identical twins" (1118). The authors conclude this subsection by telling the reader that these twin studies "indicate that genetics plays a role in determining behavior even in humans, although the relative importance of genetics versus environment is still hotly debated" (1118). This is a kind of "mentioning" as well, in which both the issue of genetic influences on humans and the sociology debate itself are "mentioned." And, most significantly, this is the only mention of a human example in the chapter. In fact, neither sociobiology nor E. O. Wilson is presented at all in the chapter. Yet, some of the key ideas that Wilson incorporated into his work on sociobiology, such as kin selection and reciprocal altruism, are discussed in detail. In the context of kin selection, both Haldane and Hamilton's work are discussed in detail (1138–1139). It is also interesting to note that the only place in the book where E. O. Wilson is discussed is in Chapter 57, titled "Dynamics of Ecosystems" (1189–1210). This is with regard to his earlier work with MacArthur on what is called the "species-area relationship" (1208), and it is work that was not directly connected to his later work on sociobiology. All of this is in stark contrast to the third edition of Raven and Johnson, wherein so much effort is spent by the authors trying to show readers why human behavior also has strong genetic influences and wherein so many rhetorical devices are used to bolster this strong genetic position.

Presentations of Genetics in the Eighth Edition

This text covers genetics in a similar manner to its coverage of evolutionary theory. There are nine excellently presented and thoroughly detailed chapters covering genetics and molecular biology. Chapter 12, titled "Patterns of Inheritance," covers Mendel's experiments in detail and derives principles of segregation and independent assortment in the process. Most importantly, Mendel's discoveries are qualified in the detailed subsection titled "12.6 Extensions to Mendel." In this section, the authors note that when scientists attempted to duplicate Mendel's research they "often had trouble obtaining the same simple ratios he had reported" (230). Furthermore, because of the traits that Mendel chose to examine, "A number of assumptions are built into Mendel's model that are oversimplifications" (ibid). Ultimately, Mendel's model is an oversimplification because it assumed that "each trait is specified by a single gene with two alternative alleles; that there are no environmental effects; and that gene products act independently" (ibid).

These issues are then taken up in the subsections that follow. In doing so, the authors point out that, among other things, "few phenotypes result from the action of only one gene" (231) and that environmental influences exert strong effects on phenotypes (233). Again in Chapter 19, titled "Cellular Mechanisms of Development," the authors take pains to point out the "environmental effects on development." What is important about the coverage of genetics is that Mendel's treatment as a canonical figure is tempered by contextualizing his work and by pointing out the lacunae in his work and in a simple Mendelianism. This is crucial because, very often, as was seen in Part I of this work, determinist arguments have tended to rely on such simple Mendelian mechanisms. These are all important qualifications that support and likely are included because of the sociobiology debate and the developments in genetics, molecular biology, and evolutionary theory that have occurred since the earlier editions were written. This qualification and the warning not to subscribe to a simplified Mendelianism are important notes to readers and help them to undermine facile arguments that connect genes to behavior through a simplified Mendelian framework.

On the other hand, there is a statement about characters and traits that is only marginally useful.

> Referring to a heritable feature as a *character*, a modern geneticist would say the alternative forms of each character were *segregating* among the progeny of a mating meaning that some offspring exhibited one form of a character (yellow seeds) and other

> offspring from the same mating exhibited a different form (green seeds). This segregation of alternative forms of a character or *trait*, provided the clue that led Gregor Mendel to his understanding of the nature of heredity (220).

This definition is only marginally useful because it is circular. What are traits? Traits are what geneticists measure. It provides the reader with no real way to think about what can or should count as a trait. Nor does this definition provide the reader with a way into thinking about traits and the ways in which they are measured that would indicate that they may have been socially constructed instead of having occurred "naturally."

Presentations of Evolutionary Theory in the Eighth Edition

As is the case with the other more recent text evaluated, increases in our knowledge about genetics and evolutionary theory are in evidence in the eighth edition. There are six chapters devoted to evolutionary theory. They cover the topics of population dynamics, the evidence for evolution, the origin of species, genome evolution, systematics and phylogeny, and the evolution of development. The text provides excellent, in-depth, clear, and thorough analyses of these topics. In Chapter 20, titled "Genes Within Populations," the reader is treated to a very thorough discussion of the influences of selection. Section 20.5, titled "Interactions Among Evolutionary Forces" (405–406); Section 20.6, titled "Maintenance of Variation" (406–407); and Section 20.7, titled "Selection Acting on Traits Affected by Multiple Genes" (408–409), all speak to the complex interactions involved in the processes that compose what we call natural selection. Also, in Section 20.9, titled "The Limits of Selection" (412), the authors cover a number of important limiting factors and remind the reader: "Although selection is the most powerful of the principal agents of genetics change, there are limits to what it can accomplish" (412). This direct reminder of the limits of selection, coupled with the excellent detail and thorough treatment of the many factors that compose the processes of selection, add up to a substantially different message about evolutionary theory than in the earlier edition of the text. The eighth edition is still a staunch defender of evolutionary theory, but the presentation is somewhat more nuanced. In my estimation it does not support an ultra-adaptationist reading and so does not directly support more determinist presentations that rely on such interpretations.

Presentations of the Nature of Science (Knowledge) in the Eighth Edition

In Chapter 1, titled "The Science of Biology," there is a section 1.2, titled "The Nature of Science," in which the authors inform the reader that scientists

> attempt to be as objective as possible in the interpretation of data and observations they have collected. Because scientists themselves are human, this is not completely possible; because science is a collective endeavor subject to scrutiny, however, it is self-correcting. Results from one person are verified by others, and if the results cannot be repeated, they are rejected. (4)

This shows a shift in position to a more contextual notion of science practice, while at the same time adhering to the idea that, as a collective endeavor with crosschecks and skepticism, consensus is reached. Later in that same section, in a subsection titled "Reductionism Breaks Larger Systems into Their Component Parts," the text tells us that while reductionism as a method and philosophical approach is a very useful tool when trying to understand complex systems, "reductionism has limits when applied to living systems" for at least two reasons (7). One is that enzymes in vivo do not always behave as they do in a isolated laboratory experiment; and,

> a larger problem is that the complex interworking of many networked functions leads to emergent properties that cannot be predicted based on the workings of the parts. Biologists are just beginning to come to grips with this problem…[and] the emerging field of systems biology is aimed toward this different approach. (7)

This notion of emergent properties references work done in systems biology in the past two decades, and it is work that seeks to develop nonreductive models of biological systems at many levels of analysis. It is a growing field with many branches and key elements of this paradigm are discussed in Kitano (2001); Klipp et al. (2005); Alberghina and Westerhoff (2005); and Noble (2006). The referencing of the systems biology approach is an important development for this book, as it is for the fourth and seventh editions of the texts by Campbell and Campbell and Reece, because, as I indicated in Chapter 2, reductionism is a keyword in the sociobiology debate. In this eighth edition, the new authors acknowledge its importance while at the same time acknowledging its limitations. That they do this through an appeal to the idea of emergent properties in biological systems is also important. In the last section of Chapter 1, titled "1.4: Unifying Themes in Biology" (12), the final unifying theme mentioned is that "Emergent properties arise from the organization of life" (14). In this subsection the authors remind the reader that "the idea that the whole is greater than the sum of its parts is true of biological systems" (14). This idea of emergent properties, the idea that the parts cannot always predict the whole, is one of the key ideas relied upon by critics who attempt to develop theories that transcend strong determinist positions, and it is a position reminiscent of Levins and Lewontin's (1985) and G. Allen's (1983) conception of dialectical development.

It is also interesting to note that Darwin is discussed in most detail in the first chapter of this edition and not in great detail in the chapters dealing directly with evolutionary theory. In this first chapter, there is a section titled "1.3: An Example of Scientific Inquiry: Darwin and Evolution" (8). As in the previous edition, Darwin is used as a canonical and archetypical figure of the best kind of science practice. It is clear and well presented and incorporates the influence of Malthus and some of the theories that preceded Darwin. This is similar to the earlier edition and as in that version, is done both as a defense of evolutionary theory and of the importance of a materialist scientific method. In this light, it is interesting to see that Darwin comes up again in detail again only in Chapter 21, titled "The Evidence for Evolution," in a section called "21.7: Darwin's Critics" (429). This section, however, is not so much about Darwin as it is a thorough and systematic refutation of seven common objections raised about the theory of evolution.

Summary of the Eighth Edition

In many ways, the eighth edition is a fundamentally different text than the third edition, and at the same time it still has, and rightly so, a staunch defense of evolutionary theory and a strong conviction of the importance of building knowledge through materialist scientific practice. The eighth edition of *Biology* puts evolutionary theory in a focal position in the text, but it does so with a more nuanced and evenhanded approach. As this textual analysis has indicated, the eighth edition does not simply mention nongenetic factors but goes to some length to outline what these can include and how they can be seen to operate. Also, the eighth edition does not use photos and inserts to slide in discussions and unqualified comments about genetic influences on human behaviors. Likewise, this text undercuts simple Mendelian and ultra-adaptationist arguments that have been used to support sociobiological discourses. And, finally, the eighth edition develops a more contextualized version of the nature of science and also introduces the idea of emergent properties in biological systems as an *alternative* conceptual framework that avoids reliance on only a reductive model of scientific investigation.

Notes

1. However, recent research appears to challenge the existence of both altruism and kin selection behavior in meercats (Clutton-Brock et al., 1999).
2. The term "heritability" has both technical and colloquial meanings. The technical, or scientific, meaning is statistical and measures the degree to which children have a biologically based, phenotypic resemblance to parents. The colloquial usage of heritability (and "heritable" as well) is synonymous with "inherited." Paul (1995) found that genetic textbooks sometimes mistakenly used the colloquial meaning of heritability rather than the statistical meaning.
3. Moody (1996) has found that, in general, biology textbooks have increased the prominence of evolutionary theory in the 1990s.
4. Michael Ruse (1996a) has chronicled a good deal of the creation/evolution controversy in *But Is It Science?* In this book he describes his role and the role of others in this debate as it took shape in Alabama courts in 1981. In 1981, creationists succeeded in getting the Alabama state legislature to consider "a bill requiring of its teachers that if they talk of the [evolution] in their classrooms, then they must talk of creation-science as well" (17).

   On March 19, 1981 the bill became law. The bill was overturned later by the courts, but creationists have continued to push for the acceptance of creationism as a scientific and not a religious doctrine. In recent incarnations, this push has seen creationism transformed into the "science of intelligent design" in the work of, for example, *Of Pandas and People: The Central Question of Biological Origins* by Percival Davis and Dean H. Kenyon (1996). In 1999, the Kansas Board of Education voted to allow the teaching of intelligent-design theory along with Darwinian evolutionary theory.
5. Unless otherwise specified, all further page references in this section and in the remaining sections of this chapter are understood to reference Losos et al., 2008.

· 6 ·

BIOLOGY: DISCOVERING LIFE

A More Qualified Presentation

This chapter analyzes *Biology: Discovering Life* (2nd edition), written by Joseph S. Levine and Kenneth R. Miller (1994). I have chosen this text because the authors were intent on covering controversial issues in biology. This is unusual in biology textbooks and in textbooks in general. This has turned out to be more than a passing interest of at least one of the authors. Miller (2008) has gone on recently to write a book on controversial issues in evolutionary theory. In this regard, this text has much to commend it. The authors are careful to qualify discussions of genetics, evolution, and behavior. This text does a good job of qualifying what counts as evolutionary theory. It also goes to great pains to connect science, technology, and society and so validate STS and STS(E) studies as an area of inquiry. The authors attempt to present controversial issues and attempt to place canonical figures in historical context. This text also directly discusses issues of the nature of science knowledge. The textbook is also unusual in other ways. It contains a chapter on "Embryology and Development." Fields such as cytology and embryology have taken a back seat to the hegemony of biomolecular genetics. Few researchers look at how an embryo forms any longer; they now look at "developmental genetics." They look at the genetic codes that seem to be dictating the development. The net effect of all these unusual discussions is that the text is comfortable with controversies in science, is comfortable with looking at science in its social context, and is comfortable examining the ways in which science activity affects the biosphere.

Presentations of Animal Behavior and Sociobiology

The last chapter of the text, Chapter 46, is titled "Animal Behavior" and, unlike the Raven and Johnson third edition, it is not constructed to give genetics an overly prominent role in the chapter's discussions. Subsections on "Genetic Influences of Behavior" and "Behavioral Genetics" make up only two of the six subsections in the "Elements of Behavior" section. The chapter is twenty pages long and in total, the discussions about behavior in the contexts of genetics, sociobiology, and evolution take up about one-third of the chapter.

Chapter 46 begins with a discussion of terminology and approaches. Six pages are then devoted to a more detailed presentation of ethology and its tenets. Included in this discussion is evidence for acquired and inherited behaviors. Again, unlike the third edition of the Raven and Johnson textbook, this book actually spends some time on ethology and on experimental evidence that demonstrates both learned and inherited behaviors in animals. In the subsection called "The Evolution of Social Behavior," the authors introduce sociobiology and the concepts of inclusive fitness, kin selection, and altruistic behavior. They present about one-half page of text on these issues in what I would call a careful way. They qualify their support for kin selection by saying that

> there is no way to prove that kin selection is universal, but a host of observations on social behavior suggest that it exists in many species.... We now know, for example, that members of many bird flocks, lion prides, elephant herds, and monkey troops are composed of close relatives—usually parents and one or more generations of their offspring. (982)

I call this discussion careful because the authors do two things. They affirm that a key concept in sociobiology seems to operate in a good many cases but also qualify their statements by disallowing any automatic claims for universality. Undermining the claims for universality takes some of the hegemonic function away from sociobiology, while allowing for it to be correct in some cases. They also specifically do not include humans in this discussion.

In Chapter 46, after the presentation of sociobiology and the discussion of the concepts of inclusive fitness, kin selection, and altruism, they provide one column of a page showing how "The most extreme cases of altruistic behavior occur among the social insects." (983). They then move on to a discussion of "Primate Societies." The opening sentence of the subsection is a qualifier. The authors say that "some of the most fascinating studies in animal behavior—and among the most difficult to interpret—involve members of our own order, the primates" (984). The section on primates is less than one column of a page and deals primarily in generalities and a brief discussion of dominance hierarchies. The authors comment that it is often difficult "not to see primate behavior in human terms." Pygmy chimps, for example, seem to "play games." The authors leave the ambiguity and issue at that level, however, and don't engage in a meaningful discussion of how one "ought" to think about primate behavior.

This subsection is followed by another subsection, titled "Human Behavior." It opens with a quote from Shakespeare, followed by another qualifying sentence:

> *Homo sapiens* is a behaviorally complex primate whose behavior concerns us a great deal. We approach the study of our own behavior from the perspectives of several disciplines: anthropology, sociology, psychology, and—of particular interest to us here—ethology and sociobiology. (986)

They approach the issue of genes and behavior directly when, in the second paragraph of this section, they ask: "As individuals, how much of what we think, feel, and do today is programmed within our DNA?" (986). In the next one half page that follows, they present a strong argument as to why one should not assume genetic causes for complex human behaviors. It is interesting to note that in this strong qualification they make reference to the first two chapters of the text, which deal with issues of science and society. They say in a discussion of the nature/nurture controversy that

> unfortunately, the history of studies into genetic influences on the human mind includes a great deal of bad science, distorted by the sorts of prejudice we discussed in Chapters 1 and 2. For it is particularly difficult for anyone to investigate links between genes and brain activity in humans without falling prey to their own preconceptions. (987)

The text then goes on to show how studies in the 1980s purporting to show that single genes control manic depression, "alcoholism, childhood depression, and schizophrenia" "began to unravel soon after they were published" (987).

This last page of the chapter contains a great deal of qualification. The authors issue a strong cautionary note concerning genetics and behavior. They say that

> it is one thing to find a single gene that affects learning in a fruit fly or genes that affect mating calls in crickets. It is entirely another to postulate (and extremely difficult—some would say impossible—to *prove*) that genes—which can code directly only for structural or regulatory proteins and not for behavior—control such difficult-to-define human traits as "intelligence," aggressiveness," and "dominance." (987)

A few paragraphs later, they discuss the difficulty of conducting research around complex human behaviors. First they say that

> human behaviors are extremely difficult to categorize neatly....This difficulty in precisely specifying phenotype—which is even more pronounced with nonpathological behavior traits such as intelligence—makes correlation with genotype extremely difficult. (987)

Yet, the next sentence sends the qualification back in the other direction: "[I]t is possible that improved diagnostic tools may someday overcome this diagnostic uncertainty" (987).

Overall, I feel that Levine and Miller present a balanced and qualified approach to the topic of animal behavior and to sociobiology itself. They present the history and findings of ethology. They discuss learned behaviors and innate behaviors. They do not shy away from the issue of genes and behavior but present sociobiology in a less hegemonic way and clearly qualify its applications, and other genetic and evolutionary applications, to higher primate and human behaviors. It is a thoughtful approach to the topic. In the summary at the end of the chapter, they reiterate their position that "in higher animals, most behaviors result from complex interactions between nature and nurture" (987–988). The final paragraph of the chapter details the kinds of experiments that cannot be performed on human subjects. The chapter, and the entire textbook itself, ends with a statement that re-establishes the link between science knowledge and society: "The answer to those experiments we can perform may someday tell us more about who we are than some of us want to know. What we as a society choose to do with the information will tell us even more" (987).

In general terms, I would say that Levine and Miller engage in a much less hegemonic exercise in their text. In the chapter on animal behavior, they make concerted attempts to present balance and evidence that support both nature and nurture. As readers will see in Chapters 8 and 9, however, the authors highlight "unusual" issues in sections called "Theories in Action" and "Current Controversy," but they include neither type of subsection in this chapter. It would seem that the nature/nurture discussion or a discussion of the controversy generated by sociobiology would be perfect topics to present as controversial issues. Yet they seem to shy away from it here but take on the very controversial issue of genetics and IQ in an earlier chapter.

Presentations of Genetics

Part III of this textbook is titled "Evolution and Mendelian Genetics" and it covers chapters 8 through 13. Chapter 10, titled "Genetics: the Science of Inheritance," begins with an interesting and important set of warnings and qualifications.

> For thousands of years, people have successfully bred plants and animals. The idea of selective breeding is simple. One chooses a few organisms with desirable characteristics to be the parents of a new generation. When that generation is grown, the hardiest plants, the fastest racehorses, or the most obedient dogs are selected for the next round of breeding. It's simple in practice and simple in theory. Offspring tend to resemble their parents, so by patiently choosing parents with the proper characteristics, the breeder can accentuate those characteristics.
>
> Is inheritance really that simple? For hundreds of years, people regarded inheritance as a *blending*—the characteristics of both parents were thought to blend to produce offspring. In most cases, the blending explanation seems to make sense. Most of us look a little like our mothers and a little like our fathers. A cross between a large dog and a small dog usually produces medium-size dogs. Everything seems to make sense. Well, almost everything. (189)

In these passages, we see an appeal to common sense similar to that made in *Understanding Biology*. In the first paragraph, there seems to be a simple and direct linking of characteristic to genetics made through the example of animal breeders.[1] There is a common sense appeal here to direct human experience. The second paragraph calls the assumptions of the first into question. In the second paragraph, inheritance is taken to be a more complex a phenomenon and common sense cannot always be trusted. At the same time, however, the analogy taken from animal breeding has stretched and now includes the human realm. The analogy appears obvious, but it eradicates the complexities involved in the processes of inheritance. The reader is then left to ponder many questions. Is inheritance that simple and straightforward, or is it much more complex? Most importantly, in these two introductory paragraphs, no mention is made of the role of environmental, cultural, or social factors in the expression of characteristics or traits. And, as I mentioned above, while these authors, to their credit, do raise important questions, I do not think that this textbook adequately answers the questions it poses to readers. Nor do I think that it provides teachers and students with the conceptual tools to begin to deal with the complexities the authors raise.

In Chapter 4 of this book, I indicated that the lack of a clear definition of the term "trait" leaves the concept open to a range of interpretations. Those advocating for a strong genetic program take advantage of this ambiguity. They feel that they have license to consider any complex behavior as qualifying for "traithood." This, in turn, allows wild speculation to abound and also tends to eradicate the complex social, cultural, and cognitive dimensions that are part of each organism's experiences. I found that this textbook has no entry in the index for "trait" or "characteristic" and only one for "character." This signals that for these authors, as for most biologists, the concept is ubiquitous and not in need of careful definition or clarification. Even though the authors actually do engage in some process of defining these terms in the body of the text, their definitions are purely descriptive and operational.

The references to characteristics and traits begin in Chapter 10 with a discussion of Mendel's work: "Mendel assumed (correctly, as it turned out) that each plant had two such units for each *trait* [author's emphasis]. Mendel called each unit a 'Merkmal,' the German word for 'character'" (192). The authors then go on in the next subsection, titled "Dominance," to say:

> One of the remarkable results of the crosses shown in Figs. 10.4 and 10.5 is the fact that plants with contrasting alleles (*P* and *p*) produce purple flowers. The purple allele of the flower-color gene is *dominant* over the white allele. (That is why purple is represented by the capital letter *P*.) Mendel carried out crosses with six other characteristics, including plant size, seed color, and seed shape (Fig. 10.6). In each case, one characteristic was "dominant" in the F1 generation. The characteristic that seemed to disappear in the F1 generation and to reappear in the F2 generation he called a *recessive* trait. Mendel explained this phenomenon by proposing that whenever a dominant allele and a recessive allele were found in the same organism, the dominant allele alone controlled the appearance of the plant. Only when two recessive alleles occurred together did the recessive characteristic or "character" emerge. (192)

Here, both character and trait are invoked but not defined except in a circular way. A character is any identifiable characteristic that is under genetic control and capable of being expressed and inherited. With this circular reasoning in place, it is then easier to "reduce" traits to genes, or at least to see traits as representations or epiphenomena of genes.

This circular reasoning and reduction of complex behaviors to simple traits, such as eye color and hair color, are also reinforced by the presentation of Mendel's work in the general discussion of genetic principles. While the terms are modern, the experiments are Mendel's. Thus, it seems to me that a direct link is made between things such as leaf color and gene loci. At the same time, a subtle link is made between "traits" encompassing ever-more-complex behaviors and the apparent reality of specific gene loci.

This kind of link is made very clearly in Chapter 21, titled "Biotechnology and Molecular Medicine," in the subsection called "The New Human Genetics." In this section, the authors discuss "the powerful techniques of molecular biology" that "make it possible to carry out a whole range of new experiments involving human genetics." They then go on to describe how the Southern blot techniques for producing gene markers on photographic film are markers of inheritable genetic material.

> After researchers worked with the bands on Southern blots for a few years, a fundamental fact began to sink in: *those bands are inherited*.... After all, the position of a band on a blot is determined by its DNA sequence—a sequence that is inherited. (445)

The next sentence invokes the tautology discussed above.

> Therefore, we can think of the bands that hybridize to a particular probe as characters, *just like red hair or blue eyes* [author's emphasis].... [S]ome individuals show *different* bands for the same probe.... A character that differs among members of a population is said to be *polymorphic* ("many-shaped"). (445)

It is clear that character has two meanings here. One defines a unique gene or DNA locus; the other defines some visible or identifiable physical trait. In this statement, the authors clearly equate specific points on DNA to characters. Indirectly, they equate genetic markers with phenotypes.

The text also includes another definition in Chapter 12. This is an operational definition of a characteristic or trait in a population that is measurable and statistically based.

> Every characteristic of organisms in a population has a frequency distribution (Fig. 12.3). That is, if you look at any heritable[2] trait in a large enough population, you will see that it has some average value and some degree of variation or deviation from the average value. (243)[3]

While it can be argued that this definition is still tautological, at the very least it provides that data taken from measurement at least conform to laws of statistical variation. Adherence to this law might at least force sociobiologists to demonstrate that a behavior that they deem to be a trait must conform to this requirement.

We can see in this treatment of traits or characteristics that the reader can be left with the impression that it is possible and indeed sound to move from consideration of "simple" Mendelian traits, such as eye color, to more complex traits. If some identifiable phenomenon or behavior displays the proper frequency in a population, then it likely has a biological basis and, ultimately, a genetic locus. Unlike *Understanding Biology* (3rd edition), however, this does not seem consistent in its overall intent and focus. I do not think that the authors wish to provide an uncritical hegemony for genetics in discussions of human behavior. Unfortunately, as I indicated in Part I of this book, the net effect of these kinds of formulations can be to create confusion among readers, to reduce a complex reality, and to create tautological arguments. All of these developments support sociobiological discourse and other biologically determinist formulations.[4]

This confusion, circularity, and reductionism can be reinforced by similarly simplistic conceptual formulations of *phenotype* and *genotype*. Levine and Miller, however, do take pains to point out that simple Mendelian dominance and recessivity are not the only ways in which genes interact and traits are produced.

> A second major contribution of Mendel's experiments was the idea that every organism has a *genetic* makeup called its *genotype*. The actual characteristics an organism exhibits are called its *phenotype*.
>
> The genotype is inherited, whereas the phenotype is produced under the influences of the environment and the genotype. (193)

In a section of Chapter 10, titled "Other Forms of Genetic Variation," they discuss incomplete dominance, co-dominance, multiple alleles, epistasis, polygeny, and contiguous variation. They spend four pages in a nineteen-page chapter discussing the ways in which genes interact to produce traits that are not situations of simple or complete dominance, or situations in which just one pair of alleles produces a specific and measurable trait. Also, the last subsection of the chapter is called "Environmental Effects on Gene Expression." This section opens with a very clear invocation to the reader: "Earlier in the chapter, when we made the distinction between genotype and phenotype, we were careful to say that the phenotype of an organism develops under the influence of its genotype *and* its environment" (215).

Unfortunately, this section is only about one-quarter of a page long. One would think that a commitment to this position would necessitate a more lengthy discussion of the idea. Also, it would be more effective if it occurred early in the chapter and more strongly framed the entire presentation. Yet, I think here that the authors simply reflect the discipline as a whole. Lip service is paid to the idea of environmental influences, but little is done to make this a bona fide position, well defended with "fact." In addition, the text maintains a clear separation between the two realms of genes and environment rather than discussing the two as elements in an inseparable process in which study of either separately is likely pointless, especially where complex behaviors are concerned.

In another chapter, however, the authors do create a more nuanced position. Chapter 11 is called "Human Genetics." Early on in this chapter, the authors say:

> In order to apply the principles of Mendelian Genetics to human beings, we must first identify an inherited character that is controlled by a gene. This is not easy for it is often difficult to determine which characters are directly inherited and which are related to environmental influences. (218)

The importance of this statement is that it delineates a difference between human subjects and other living creatures in terms of complexity. It also preserves a role for environmental issues while highlighting the inseparability of the two kinds of influences. It still maintains a rigid duality and the possibility of separating and quantifying the different influences, however. As I discuss in Part I of this book, such a separation obliterates the actual lived reality of the individual or organism. To drive this point home, the authors also go on to say: "[M]ore than 4,000 human genes and their alleles have been mapped to specific chromosomes....At least 1,000 known human traits are produced by recessive autosomal alleles" (219).

They go on to describe rather straightforward Mendelian conditions, such as albinism, Tay-Sachs disease, cystic fibrosis, galactosemia, and sickle-cell anemia, as well as autosomal dominant conditions, such as Darwin's tubercle, achondroplasia, Huntington's disease, and polydactyly. They also go on to describe medical conditions in which multiple alleles are involved and polygenic conditions exist. Ultimately, this presentation again leaves the reader with two very different ideas. One is that we must be careful in our considerations of the relationships between genotype and phenotype in humans. The other is that we already have made a great number of direct genotype/phenotype connections in humans and that this knowledge will soon increase greatly. So, despite the authors' claim as to the difficulty in applying simple Mendelian genetics to humans, the reader is shown that this is indeed already partly accomplished.

The issue of a definition of what should be considered a trait or character is not clarified any further in this chapter. And, in the midst of discussions of traits that are mostly abnormal medical conditions, the authors insert a "Current Controversies" page with a topic titled "Human Intelligence: Are There Genes for IQ?" (224). This is an interesting placement, because it comes within a larger discussion of the linking of physical and medical traits with relatively simple Mendelian genetic models. It therefore allows for the possibility or suggestion that the reader consider IQ to be a genetic trait or character.

The controversy is described by the authors:

> Just as individuals differ in physical characteristics, they differ in mental characteristics too....Intelligence quotient scores differ widely among the human population....However, the claim that IQ tests measure intelligence has been challenged by many investigators, some of whom have charged that the tests are flawed by hidden racial and ethnic bias. *We will leave those questions unanswered and instead will ask a question that is a bit more biological* [author's emphasis]. Are individual differences in IQ scores the result of genetic differences between individuals or the result of differences in the individual environments that children experience in their formative years? In short, are there genes for IQ? (224)

From here the authors go on to describe the controversy around the lower average IQ scores of American blacks. They cite Jensen and Shockley's claims that "differences in average academic performances between white and black children are biological and cannot be eliminated by improved schooling" (224). They go on to conclude: "There is substantial evidence that the ability to score well on an IQ test may indeed be inherited to some degree" (224). And they go on further to cite evidence from identical twin studies supportive of this position. Then they qualify this statement by saying that

> the situation is not as simple as these statistics may make it appear. Studies of children adopted from orphanages have shown that adoption itself may raise the adopted child's IQ score by as much as 10 points, and there is a very high correlation of IQ level with social and economic status. Individual IQ is now known to be variable, which is to say that it can be changed by study and a positive learning environment, and it may be dramatically affected by self-image....At this point, there is little doubt that intelligence is shaped by both genetic and environmental factors, but there is little scientific support for the notion that biological differences will undermine the positive effects of improved schooling. (224)

The text then goes into great detail in the next section, describing the relationship between genotype and fingerprints, including a table that provides data on the "Correlation Between Relatives for Total Dermal Ridge Count" (225). From there, the chapter discusses the connection between one's genotype and a variety of human traits, such as skin color, hemophilia, color blindness, and a variety of medical conditions. It further presents information about sex linkage, chromosome deletions, and translocation.

Toward the end of the chapter is another "Current Controversy." This one is titled "From Mom or Dad? Gene Imprinting." This insert details the fact that the same allele donated from a male or female parent can have different effects on phenotype. They use the examples of Angelman syndrome and Prader-Willi syndrome. Those who inherit the problem gene from their mothers have Prader-Willi syndrome, while those who inherit it from their father develop Angelman syndrome. The insert shows the reader that there seems to be a phenomenon called "Genetic Imprinting" that allows the embryo to distinguish the origin of the parental contribution in ways that affect development of the embryo. To me, this does not seem like a controversy in the way that a discussion of genes and IQ is a controversy. Rather, it seems that this item is placed here to give the student a sense of the kinds of complexity with which geneticists must contend. This is an important point, precisely because critics have long held that one of sociobiology's problems is its simplistic application of genetic principles to behavioral studies.

On the last page of Chapter 11 is a brief discussion of "The Human Genome Project." It is only two paragraphs long and ends with the statement:

> Rapid progress on human gene mapping is to be expected in the years ahead, and it may not be too long before we can say with complete honesty, that we know just as much about genetics of *Homo sapiens* as we do of *Drosophila Melanogaster*. (238)

While the authors have taken some pains in the chapters on genetics and sociobiology to distinguish human genetics and its complexities from the genetics of other life forms, in the end they indicate to students that soon the level of information about both realms will be equivalent. What is unsaid here, but possibly conveyed to students, is that there is another kind of equivalence between our understanding of the behaviors of fruit flies and people. This kind of association has a way of undermining the idea that human biology and human behavior must involve complex and nonreductive explanations. Instead, this kind of association subtly lends credence to the idea that there can be a straightforward, one-to-one correspondence between genotype and phenotype, and between gene and characteristic, trait or behavior.

Chapter 20 is devoted to "Molecules and Genes." This is significant and acknowledges the important weight that molecular genetics now has in biology. In this chapter, they present information on the structure of nucleic acids. Chapter 21 is called "Biotechnology and Molecular Medicine."[5] In Chapter 22, in a "Theory in Action" insert called "Cystic Fibrosis: A Triumph and a Warning," the authors conclude:

> The Human Genome Project will ultimately identify a host of genes that control traits. Should a couple be able to screen its fertilized eggs and eliminate those that carry *any* defective genes? Should they be able to opt for a fetus with a genetic makeup that produces the sex, height, weight, and hair color they want? And critically, who has the power to make that decision? The time is not far off when we will be able, theoretically, to redefine the human condition by manipulating our genes. We should consider the implication of this power sooner rather than later. (450)

This is an interesting passage. It is important to problematize the social dimension of science research, and the authors clearly want to do that here. They raise the specter of eugenic uses for biotechnologies that may emerge from the Human Genome Project. Yet they do not present any of the issues that critics have cited about the claims the project has made.[6] Critics of the Human Genome Project have predicted that it would not provide what people expected from it, that much of the code is nonsense code, and that even with markers mapped, there could be no necessary, simple extrapolation from these markers to complex human behaviors. The authors present only supportive examples such as the "triumph" of finding the gene abnormality that causes cystic fibrosis.

Again, here we see the same kind of qualifying that I discussed earlier in relation to other controversial topics. It seems to me that while these kinds of examples begin to raise important issues and qualifications in the genes/behavior debates, in this text they are only developed in part. This incomplete development will likely lead only to confusion for students. They have just read a great deal of information that links traits of simpler creatures with Mendelian genetics. They also have received information that links traits or characters of humans with relatively simple Mendelian genetics. Throughout this process, they are admonished to remember the complexity involved for humans and to remember the role of environmental influences. Unfortunately, in the end they receive no real practical data or concepts to help them exercise this caution.

Presentations of Evolutionary Theory

Chapters 8, 12, 13, and 22 deal with topics related to evolutionary theory. This text is unlike Wilsonian and Dawkinsian texts in its avoidance of confusing talk about genes as actors in evolution. Chapter 12 is titled "Darwinian Theory Evolves." The text looks at western worldviews before Darwin, theories such as uniformitarianism, catastrophism, and Lamarckian theory. In the discussion of Lamarck the authors make an important point about its teleological nature. According to Levine and Miller, Lamarck's "theory was teleological; he believed that evolution had a goal, or directed purpose, and that species changed over time because they 'wanted' to 'better' themselves" (152). In the text's presentation of Darwin, there is no overly adaptationist formulation of evolutionary theory. For example, in the pages that discuss the development of Darwin's evidence and theory, the authors make the following statement about his notion of fitness.

> Throughout his journey, Darwin marveled at the "perfection of structure" that made it possible for organisms to do whatever they needed to do to stay alive and produce offspring. Darwin called this perfection of structure *fitness*, by which he meant the combination of all traits, physical and behavioral, that help organisms survive and reproduce in their environment. (155)

In the subsection called "Observable Variation in Organisms," the authors point out:

> Remember that genetic variation is random. It does not occur because an organism *needs* or *wants* to evolve.... In other words, alleles do not "know" precisely how to mutate or recombine in order to benefit their organism. (243)

This is important information. Unfortunately, it is presented separately from the discussions on evolution and behavior presented later in the book. These authors make a point of avoiding the kind of confusion and doublespeak that Wilsonian sociobiological discourse uses to support its arguments. There are no selfish genes or robotic organisms in this text. All in all, this chapter is quite balanced. It presents a warning about teleological formulations of evolution, and it presents a complex look at our current understandings about evolution. This is best summed up at the end of the chapter:

> Evolutionary biology today, which has absorbed significant contributions from Mendelian genetics, population genetics, and ecology, deals both with the facts of evo-

> lutionary change and with theories about how and why evolution takes place.
>
> The gradual accumulation of evolutionary change within populations can often be explained by the differential action of selective pressures on different phenotypes that represent different genotypes. Recent advances in molecular biology have revealed unexpected riches of genetic variation in natural populations. Some of these variations are sufficient to provide the raw material necessary for evolution to operate; others may be neutral, or selectively meaningless.
>
> Natural selection is not necessarily responsible for all evolutionary change. Additional mechanisms, including genetic drift, operate more according to chance than in terms of selective advantage. (257)

Yet there are still ideas missing from this presentation. These include a discussion of selection with a cooperative rather than competitive engine and, the effects of serendipity as Gould (1989) has described them. To their credit, the authors do allude to these issues briefly in Chapter 13, titled "Evolution of Species." At the beginning of a section titled "Current Debate on Evolutionary Theory," they say:

> Recent advances in genetics, molecular biology, and paleontology have called into question certain assumptions about evolutionary change. Is natural selection the only force behind genetics change? Are all evolutionary changes adaptive? Does evolution really have to progress slowly and gradually? *Spirited debates on these issues continue today* [author's emphasis]. (269)

From this point of departure, the authors discuss Gould and Eldredge's challenge of the idea of gradualism in the transformation of species and their proposal "that evolution proceeds in a manner they call *punctuated equilibrium*" (271). They cite the fact that Gould and Eldredge developed this idea after a careful scrutiny of the contemporary fossil record. In so doing, they at least present evolutionary theory as a dynamic and malleable theory that is to some degree subject to revision and modification. Inclusion of this material signals that the authors are aware of the larger debates that have been ongoing and that have an impact on the credibility of sociobiology.

Chapter 22 is called "Molecular Evolution." It features a very fascinating discussion of the mechanics of evolution by mutation at the molecular level. It shows the power of molecular understandings of genetics for both genetics and evolutionary theory. In a useful discussion of hemoglobin-S, the type of hemoglobin produced by people recessive for the sickle-cell trait, the authors show how this recessive gene conferred immunity to malaria and how it spread through populations chronically exposed to the malaria parasite. In this discussion, the authors once again bring up an important point about evolutionary change.

> Hemoglobin-S was not an ideal solution to a problem that was "designed" with any sort of forethought; it was a random mutation, no more, no less. It just happened to make life difficult for the *Plasmodium* parasite without interfering too badly with hemoglobin's vital function—*in heterozygotes*. (467)

Likewise, in this chapter there are discussions of the makeup of the human genome. The authors point out that "It now seems clear to many researchers that the entire human genome was duplicated once, and then portions of it have been duplicated several times since then" (468). And, in a "Current Controversies" insert in Chapter 22, titled "Junk DNA," the authors show that as much as 95 percent of the human genome does not appear to code for proteins and is labeled as "junk." They hope students take a lesson from this finding.

> These findings should remind us that we still know very little about how genes coordinate and control their activities....Simply because significant parts of the genome don't code for recognizable pieces of proteins doesn't mean that they don't do anything important....Garbage is something we throw away altogether....Junk is something we keep around because it might just come in handy some day. (473)

The last subsection of Chapter 22 is called "Genetic Differences and Human Populations." In this section, the authors tell the reader that "genetic differences among human populations can be invaluable research tools. . . ." (474). Yet they also warn that:

> Unfortunately, misunderstandings about the nature and significance of human genetic diversity have caused a great deal of pain and suffering throughout human history. There has been a resurgence in recent years of terms such as *racial purity* and *ethnic cleansing*. Molecular studies of human genetics, however, show that these terms, in addition to being morally repugnant, are scientifically meaningless.
>
> Molecular analyses of human DNA proves, for example, that alleles related to malaria resistance make a mockery of the concepts of "race" and "ethnicity."
>
> On a broader scale, however, the differences among humans that we can *see* (and thus usually think of) as human genetic diversity—the sorts of cosmetic variations that are often used to distinguish among so-called "races"—are caused by no more than 6 percent of the *total* genetic variation our species possesses. In other words, only 6 percent of the *total* human genetic diversity identified to date separates along those so-called racial lines. In similar fashion, only about 8 percent of our total genetic variation is divided among the more numerous and smaller populations that we often refer to

> as ethnic groups or nationalities....Roughly 85 percent of human genetic diversity occurs among people *within* any single population. (474)

Yet by contrast, if the authors are so convinced that race is not an important biological category, then why do they not bring up these points when they discussed IQ and intelligence in a "Current Controversy"? It is this lack of follow-through on points of controversy that leads me to believe that, in the end, this text is likely to leave the reader more confused than it is to give the reader tools and information to create informed opinions about these controversies.

Presentations of the Nature of Science (Knowledge)

This text seems to have a decided and clearly demarcated science and technology studies orientation and, for its time, a progressive and somewhat constructionist presentation of the nature of science. The detailed table of contents highlights, in many chapters, sections entitled "Theory in Action" and "Current Controversies" that are relevant to the specific chapter topics. For example, in Chapter 11, entitled "Human Genetics," the "Theory in Action" segments are titled "Genetic Counseling: Knowing the Odds," "Prenatal Genetics," and "What Turns off the X." The "Current Controversies" sections are entitled "Human Intelligence: Are There Genes for IQ?" and "From Mom or Dad? Gene Imprinting."

The authors state "that a book's opening chapters should enable students to appreciate the procedural and historical background of biology" (xxvii). In this regard, Chapter 1, entitled "Understanding Life: A Crucial Responsibility," contains not only the usual discussion of the scientific method but also contains a subsection called "The Nature of Science." This subsection, in turn, has subheadings that discuss "Science, Myth, and Religion"; "Science as a Way of Knowing"; and "Science as a Mirror of Society." This is unusual. Most textbooks do not usually enter the discussions of science as a social product.

The general theme of locating science as a social activity carries forward to Chapter 2, titled "Science and Society." This chapter opens with a moving account of a small town coming to grips with the long-term effects of deadly chemical pollutants in the pasture where its children played. And even though one of the parents was a biologist, she "had not given environmental issues much thought" (19). The chapter encourages the reader to make links between science and society. One topic discusses "Society, Disease, and Medicine: In Time of Plague." Another discusses "Ecology and Economics: Tending Our Houses." And although these first two chapters are short, only 29 pages in a book of 988 pages, they do set the tone that is carried forward in many parts of the text.

While Levine and Miller do take some time in Chapters 1 and 2 to discuss social influences on science, scientists, and science fact and theory, they ultimately take a conservative position with respect to the ways in which context, self-interest, and power influence science knowledge. One of the "Theory in Action" inserts, titled "Nettie Stevens and the Y Chromosome," begins with the following paragraph: "The History of Genetics, like that of any science, is filled with wrong turns, mistakes, and errors in judgment that time and the experimental method have corrected" (207). The net effect of this orientation is to lead the reader to conclude that, although there are social influences on scientists and the work they do, in the end, the method of science ensures truth—or at least accuracy. This theme will come up again in the discussions on Darwin.

In relation to the treatment of canonical figures with respect to the presentation of the nature of science, the interesting treatment in this text is that of Darwin. Chapter 8 is titled "Darwin's Dilemma: The Birth of Evolutionary Theory." Chapter 8 is interesting in that it begins with a discussion of the historical context of Darwin's voyages and his theory. Malthus is presented not as having influenced the form and content of the theory of evolution but as having been a trigger for Darwin.

> Darwin documented his growing belief in evolution and his belief and his search for a mechanism that could explain it. Then in October of 1838, he read "for amusement" a 40-year-old essay that crystallized his thinking, the *Essays on the Principle of Population* by Thomas Malthus. (156)

There is detailed discussion in Chapters 5 and 7 about Malthus' work and the controversial issues around global population growth. The authors also acknowledge the influence of breeders on Darwin:

> Darwin built his argument brilliantly. His first chapter, "Variation Under Domestication," details how farmers take advantage of random variation in crops and livestock. In a process called artificial selection (intuitive selective breeding) farmers choose the most desirable cows, sheep, or tomato and corn plants for breeding. Over several generations, this selective process produces individuals that differ markedly from their forebears. In Darwin's words, "The key is man's power of accumulative selection:

> nature gives him successive variations; man adds them up in certain directions useful to him." (158–159)

Here, the authors seem to present evidence that supports some of the claims by people such as Robert Young (1985), who argue that the shape of Darwin's work was significantly influenced by his sociocultural and economic context. They do not allow, however, that the theory of evolution could have been influenced by social context toward one form of theory over another. And, in the next subsection, titled "Natural Selection," they continue:

> Darwin's next insight was to see in nature an analogue to the farmer as selective agent....Darwin searched for a direct material force, or scientific mechanism, to drive evolution. Darwin called that material force *natural selection* and described it as a process that favors the survival and reproduction of those organisms exhibiting variations best suited to their environment. (159)

The authors understand the problem of teleology, but do not apply it at all to Darwin's work. The authors allow no ambivalence for Darwin's theory. According to them,

> although Darwin compared natural selection and artificial selection, he emphasized that natural selection operates without the foresight and purpose of the farmer. Darwinian evolution is *nonteleological*, which means that the process of natural selection operates without any ultimate goal of perfection. Natural selection operates only in the here and now for each organism. (160)

This is an interesting reading back into history. The authors are imposing their own understanding of evolution back onto Darwin and, at the same time, Darwin's stature as an icon in biology lends credibility to the authors' position. To their credit, the authors do give Darwin and his theoretical work some context for its development. They ultimately maintain, however, a conservative position on science as a social process. Outside influences are allowed to affect the development of Darwin's theory only to the extent that they act as catalysts that provide metaphors that lead Darwin to the "truth." The idea that Darwinism itself could contain problems and ambiguities can only be allowed to the extent that Darwin lacked a theory of heredity (171).

Also, unlike the third edition of *Understanding Biology*, in which the authors directly address the issue of creationism, in this text the issue is left unspoken. There is a clear subtext in this chapter that seems to speak to creationism without actually naming the movement. Given the ways in which this text seems unafraid to present controversial issues in biology, it is an interesting editorial decision to leave creationism out of this discussion of evolutionary theory. It would have been an opportunity to clarify the ways in which Darwin hoped to create a materialist understanding of evolution as opposed to a theological understanding.[7]

Summary

As I indicated in the beginning of this chapter, this textbook attempts to discuss many unusual and controversial topics. It is impressive that the authors have taken this risk and have attempted to present students with some sense of the dynamic, changing, and social process that is science. The text does not fully achieve this goal, however, and appears to fail to provide the reader with the necessary information to create informed opinions about the controversies it covers. For example, the authors of this textbook go some distance to try to elucidate the complexities and controversies that arise with any discussion of genetics and behavior. This text seems to walk a fine line, however, between a presentation of genes as controlling most traits, and even human behaviors, and a presentation that seeks to sensitize the reader to the dangers of a simplistic reductionism. Their presentations on these issues are not accompanied by a clear framework on how to think about the relationships between biology and behavior, about the nature of biology and the organisms it studies, and about the nature of science itself.

This lack of clarity suggests that their presentations may serve more to confuse than enlighten the reader. They take the reader to a place of questioning but provide no framework to answer those questions in a nontraditional way. This text does not go far enough in laying out clear opposing positions for controversial issues in sociobiology. When coupled with the confusion such incomplete presentations can generate, it is likely that, in attempting to overcome this confusion, students may either ignore the issues or fall back on more traditional formulations that permeate popular discourse and common sense. By this I mean interpretations that reinforce the hegemonic positions that emphasize a combination of basic Mendelian principles and link these to complex behaviors as traits that are subject to selection pressures. Also, though the text frames the social context of scientific discovery, in the end it falls back on a modified neopositivist version of the nature of science and science knowledge.

Notes

1. As I indicated in Chapter 2, Young (1985) has made the observation that Darwin was unduly influenced by the work of animal breeders and this encouraged him to assign selection an over-determining role in the process of evolution.
2. See Note 2 in Chapter 5.
3. While it can be argued that this operational definition is still tautological in character, at the very least it provides that data taken from measurement conform to laws of statistical variation. Adherence to this law might at least force genetic reductionists to demonstrate that a behavior they choose to call a trait can be quantified in this way.
4. I found an interesting piece of anecdotal evidence to support this hypothesis. I came across a blog from a student at Bryn Mawr who was writing about an early-1990s edition of this textbook. The student writes that she "really appreciated the emphasis on science as a perpetually changing way of thinking and looking at the world.... This attitude gives students motivation to keep exploring and questioning.... Unfortunately, the book is self-contradictory; 99% of this book presents a different view of science" (Mellors, Dec. 21, 2006). I found this blog at http://serendip.brynmawr.edu/exchange/node87.
5. I will not go into detail although I think that the issues around new biotechnologies are far-reaching and very important. These issues need space in biology texts, but that research will have to wait for another study.
6. See, for example, E. Keller, 2000.
7. There is evidence that Darwin did not succeed in this task, however. Young (1985), in his book *Darwin's Metaphor*, argues that, while Darwin strove to create a science of evolution based on suppositions grounded in material observations, his work was also influenced by the natural theology of the time and by his observations of the work of animal breeders. The net effect, Young feels, is that Darwin's work is ambiguous and contradictory to the extent that the concept of selection is sometimes imbued with an intentionality borrowed from natural theology and from the example of animal breeders.

· 7 ·

BIOLOGY: PRINCIPLES, PATTERNS, AND PROCESSES

A Canadian Compromise

Biology: Principles, Patterns, and Processes was the only Canadian textbook on Circular 14, the Ministry of Education's list of biology textbooks approved for use in Ontario Academic Courses (OAC) in biology in Grade 13 in Ontario high schools.[1] (See Appendix I for the complete list.) This is *Biology: Principles, Patterns, and Processes*. In general, it is less "flashy" than its American counterparts. There is less use of color and specialized highlights. It is also older than the other books I have reviewed, but it is the most recent Canadian text that covers introductory biology at this level and as extensively as the other texts reviewed.

This text, more than the others, is a joint effort with many contributors and with Galbraith acting as much as an editor as an author. For this reason, it is likely that this text represents more of an attempt at compromise than the others, and it contains "equal time" for various controversial and contradictory positions. Yet, as I indicated with my analysis of *Biology: Discovering Life* (Levine and Miller, 1994), this approach may lead only to confusion in students or, worse, may lead them to ignore the issues entirely. In the end, these reactions likely will leave the legitimacy and credibility of determinist formulations intact.

Presentations of Animal Behavior and Sociobiology

I found the presentation of sociobiology and the general presentations around biology and behavior disappointing. The authors seem to start off their presentation by strongly qualifying biological claims on animal and human behavior but end up seeming to do the opposite. Chapters 23 and 24 are devoted to animal behavior. Chapter 23, "The Behavior of Animals," begins by saying that

> in one way or another, the behavior of animals in their natural habitats is usually *adaptive*....In normal circumstances, an animal's behavior usually serves to maximize its chances of survival and/or reproduction. In short, an individual's behavior increases its *fitness*. As you will see in the following sections, the concept of fitness can provide a perspective from which to view the behavior of all organisms. (450)

The text then goes on to discuss birdsong behavior and to present it within the general framework of stimulus response behavior. It discusses hormonal fluctuations and rhythmic behaviors, such as circadian rhythms (450–453). Next, there is a section titled "The Behavior of Simple Animals" that looks at the nervous systems of insects and earthworms. It presents concepts such as kinesis, taxes, and reflexes. The chapter ends with a presentation of the "knee-jerk" reflex in humans, thus creating a segue to the next section, titled "The Behavior of Complex Animals." In this section, innate inherited behaviors are contrasted with learned behaviors. In this presentation, more text is devoted to innate behavior. At the end of the discussion of learned behaviors, however, the author makes an important point by saying:

> Learning is particularly important for animals that live in groups. In such environments, relationships among individuals can determine who will get a mate, who can feed, who will get the best den, or any number of other traits. Since these relationships are constantly changing, only if animals are able to learn new associations rapidly can they continue to behave appropriately in various situations. (461)

What I find useful in this statement is that it makes a link between learning and complex social behaviors and does not try to reduce these complex situations and behaviors to genetic dictates. What I find problematic is the use of the word "traits." As I have indicted previously in Part I of this book, there is often no clear definition for "trait" or "characteristic" in biology. Most quasi-definitions I find are circular references. This lack of clarity allows anyone to claim any "thing" or "behavior" as a trait. Sociobiologists and behavioral geneticists do this all the time. In this case, the author seems to be saying that "who can feed," "who will get a mate," and "who will get the best den" are traits, presumably as eye or coat color are traits. But former outcomes are, as the authors indicate, the product of relationships in complex settings. How can they be traits unless there is a supposition of some correspondence between a phenotypic set that creates the conditions for the outcome and is determined by a genotypic set?

This theme of compromise and multifactorial modeling continues in the subsection called "The Nature-Nurture Controversy." The authors take the position that it is now a defunct controversy because

> this debate has been largely resolved; most explanations of behavior involve a compromise between these two views, allowing that the behavior of almost every type of higher animal incorporates *both* learned *and* innate components. As you will see in [Chapter 24], much of the original controversy stemmed from the fact that researchers initially approached the study of animal behavior from two distinctly different perspectives and employed contrasting methods of study. As a result, they produced what were often conflicting interpretations of animal responses. Yet both kinds of interpretations have contributed substantially to our knowledge of the functions and mechanisms of behavior. (461)

The rest of the chapter involves subsection 23.4 and a discussion of what is called the "Behavioral Continuum." This section covers the continuum of behaviors that range from innate to insightful. It includes discussions of closed instincts, open instincts, habituation, restricted learning, imprinting, flexible learning, exploratory learning, trial and error learning, and insight learning (462–469). Only in the case of insight learning are primates and human children used as examples. For the rest of the types of learning, only "lower" animals are used.

The one example of birds engaging in insight learning, however, is given in Figure 23.31 at the end of the chapter. The picture shows a bird with its beak stuck through the top of a milk bottle. The caption reads:

> For many years, English milkmen had been leaving their deliveries outside, on porches and front steps. In 1951, a few individual great tits (a species of bird found in England, which is related to the North American chickadees) suddenly learned how to open the milk bottles to drink from the top. Within a short period of time, many members of the species were performing this "trick." The behavior became a tradition that people now have to guard against. (469)

The importance of this example is to show that even among "lower" animals such as birds, insight as well as learning can occur and can be taught to other members of the population. Sociobiologists like to separate humans from other animals in allowing little for environmental influences on animal behaviors while, at least formally, allowing for environmental influences on human behavior. The more it can be shown that other animals engage in thought and learning shaped by experience, the less landscape sociobiology will be able to claim under its genetic hegemony.

Chapter 24 is called "The Study of Animal Behavior." It begins with a discussion of the dangers of anthropomorphism on page 473. The authors warn that

> modern researchers attempt to avoid anthropomorphism, the attribution of human characteristics (including emotions) to animals.... It is anthropomorphic to say, for example, that a scolded dog slinks onto a corner because it is "ashamed," or that a bird sings because it is "happy." (473)
>
> In fact, anthropomorphic interpretations of animal behavior may be quite wrong.... The causes and effects of much human behavior *may* be similar to those of other species. It is nevertheless more productive to avoid automatically assuming that this is so.... Care must be taken to avoid biasing the study with human prejudices. (474)

A discussion follows of the variation in senses and ranges of stimuli that can be detected for living creatures. The authors reiterate the dangers of anthropomorphism when they show that

> even animals more closely related to *Homo sapiens* (those whose behavior is most likely to be explained anthropomorphically) have quite different sensory worlds (Figure 24.4). For example, humans are quite visually oriented.... Dogs, on the other hand, are color blind.... Their activities are largely determined by their sense of smell. (476)

The text flows quite nicely from there to a discussion of ethology and the major ethologists. It presents some of the "fixed action patterns" and "Innate Releasing Mechanisms" that have been studied. It also includes a subsection on "Innate Behavior and Learning," in which the authors counsel the reader that

> upon closer inspection, many so-called fixed action patterns turned out to be less stereotyped than originally thought.... A male stickleback does not court *every* appropriate female, nor does every female he approaches respond. Although scientists are not completely certain how mate selection is made, it involves past experience—learning—to some extent. (482–483)

This is a very important passage because it clearly says that even in "less complex" creatures with less complex nervous systems, it is inaccurate to talk about fixed innate behaviors without also talking about experience and learning. The authors seem to reverse themselves, however, in the discussion that occurs later in the chapter, concerning sociobiology.

The section that follows "The Ethologists" is called "The Behaviorists." It focuses on the various forms that learning experiments with animals have taken. It also includes a subsection titled "Learning and Genetic Programming." Much in the same way that they tempered the ethological presentation of innate behaviors, the authors temper the behaviorists' presentation of learned behaviors. They end the section on behaviorism with a warning:

> [M]ost animals have strong innate (and therefore species-specific) biases that limit their learning abilities in ways which are beneficial to them....[F]ew animal behaviors are entirely without an innate component; the majority are influenced, even if only subtly, by genes. Just as very little of animal behavior is wholly innate, very little (if any) is entirely learned. (487)

A two-page insert follows the discussion of behaviorism, titled "The Science of Animal Behavior" (488–489). The insert attests to the fact that

> like researchers in other fields of science, students of animal behavior employ the scientific method. In order to answer questions, they formulate *hypotheses* and collect *data* that will test predictions arising from these hypotheses. They then analyze the data objectively, usually with the use of statistical tests....(488)

The insert goes on to present a possible set of hypotheses and experiments that could be run to test them. Data are presented, and the reader is told that researchers would then test "their results with appropriate statistical tests that allowed them to determine *objectively* (without bias) whether the results agreed with the prediction or not" (489).

This insert appears to come out of the blue. There is no discussion preceding it that attempts to justify the method of science or legitimize the work of people studying animal behavior. It is clear in this insert that the scientific method is being used to legitimize the study of animal behavior. It is interesting to note the placement of this insert in the chapter. It does not come at the beginning because the sections on ethology and behaviorism are presented as historical information. It comes before the section titled "Studying Animal Behavior Today," which also is almost entirely a discussion of sociobiology. It is as if the authors felt the need to remind the reader that if one follows the *appropriate* scientific procedure, one can do research on animal behavior. Of all the "Special Features Boxes," as they are called in the text, this one and another one five pages farther on (pages 494 and 495 in the same chapter) are the only ones that deal with the epistemology, process, and technique of doing science. And it is interesting that this discussion takes place in the context of research on animal behavior. In my discussion of sociobiology in Chapters 1 and 2, I said that sociobiology relies heavily on the legitimacy of the "proper" scientific method to legitimize its own area of study and claims. This discussion, coming where it does in this chapter, seems to mark a turning point in the author's emphases.

In the first paragraph of the section entitled "Studying Animal Behavior Today," the authors say:

> Using the highly technical instruments and methods of *neurobiology*, researchers have been able to identify the parts of the brain (in some cases, even the specific neurons) involved in the performance of certain behaviors (Figure 24.22).
>
> *Behavioral Genetics* has used the mathematics of population genetics and the laboratory techniques of molecular genetics to demonstrate the importance of heredity to the performance of certain behaviors. In a few relatively simple organisms, the specific genes coding for a particular response have been isolated (Figure 24.23). (490)

Figure 24.23, referred to in the quote above, describes an experiment in which a fruit fly "learns" to associate an odor with an electric shock. The authors go on to observe that "flies with a single mutant allele on an identified gene cannot learn this task" (490). Flies with another mutant allele "learn the task normally, but forget their training unusually rapidly" (290). The page also presents the concept of *behavioral ecology*, which they describe as "the study of the relationship between an organism's behavior and its ecology; that is, how both the physical and the social environment of a species affect the behavior typical of its members" (490). The net effect of all this information seems to be that pages 488 through 490 of this chapter take the reader from a position of balance to a position of extremes. In the position of balance, both genetics and environment contribute to learning. In the extreme position, the "science" of animal behavior is beginning to locate the genetic sites that control learning and memory.

The next section is titled "Sociobiology." According to the authors, sociobiology is concerned with the biology of the social behavior of animals: "The term "social" here implies *both* aggressive *and* cooperative behaviors" (491); "A major objective of sociobiology is to uncover the role of natural selection in shaping social behavior" (491). The chapter devotes almost four pages out of twenty-five to the discussion of sociobiology. It discusses the concepts of fitness, inclusive fitness, kin selection, and altruism, and, importantly, presents a subsection titled "The Sociobiology Debate." In this subsection, the authors point out the differences between biological determinists and environmental determinists. It is interesting to note that, in this chapter, no examples of environmental determinism are given, but the archetype is resurrected here so as to allow the text to claim a middle ground between the two. This is reminiscent of the discussion I presented in Chapter 2 on what is called "the propaganda of the middle ground." The text ends up taking a seemingly middle-ground position and ending the chapter with this passage:

> Most scientists, including most sociobiologists, do not argue for either complete biological or complete environmental determinism. The real truth, as usual, probably lies somewhere in between.…Human social behavior should be viewed as the result of a complex interaction of genes and environment. Much can probably be learned from attempts to discover the extent to which each of these factors affects human responses.…Although there is no guarantee that it will provide these benefits, the sociobiological approach is probably worth investigating. It could provide a more complete—and potentially very useful—understanding of why humans behave as they do. (496)

It seems here that, ultimately, the authors end up telling the reader that human sociobiology is essentially correct as it is framed. What is also interesting here is that the discussion of sociobiology itself is confined exclusively to nonhuman examples. The section on the sociobiology debate is exclusively about sociobiology and people, however. There is no attempt in the chapter to delineate how and when sociobiology may be more or less relevant. It is, on balance, a favorable account for sociobiology on all levels. It would appear to me to leave students and teachers confused. First, one is told that genetic and environmental factors are equally important. But in the end, the qualifications lose a great deal of their weight; as the pages go on, there is a slow and steady shift toward considering the importance of genetic causes for both human and nonhuman behaviors.

Presentations of Genetics

In this textbook, Unit 5 is devoted to "Genetics: The Nature and Expression of Genetic Information." Chapter 14 begins with a discussion of the importance of Mendel's work, describing his experiments in detail. In the context of the discussion of Mendel's work with true-breeding tall and dwarf plants, the authors explain that the reappearance of dwarf plants in the F2 generation "Suggested that the character was determined by a *unit factor* that must have been present unaltered in the hybrid F1 progeny" (270); F2 refers to the offspring from the second pairing and F1 refers to offspring of the first pairing. There are no entries for "characters" or "traits" in the index of the book. This is the first mention of character in the text and it comes without explanation or definition. The text goes on to explain that "here was a clear statement of what was later to be called the *concept of the gene*. The name *gene* would only later be applied, but the concept of discrete units of heredity had been clearly stated" (270). In these passages, one can see that, for Mendel, "characters," such as plant height, are being constructed as "traits" and connected in a "determined" way to genetic sites.

The authors then take pains to modify this early formulation. The text goes on to cover ideas such as homozygous and heterozygous alleles and dominant and recessive alleles, and the terms "phenotype" and "genotype." It also presents ideas of segregation, independent assortment and mono-, dihybrid, and test crosses. In addition, it discusses chromosomes, haploidy, diploidy, mitosis and meiosis, autosomes, and sex chromosomes. Most importantly, page 271 of the text contains an insert called "Heredity and Environment." In it the authors discuss the complexity of gene expression. They explain: "Two individuals, both with the same genotype for a given gene, often show some variation in their phenotypes."

The insert goes on to explain that "the gene for pattern baldness shows greater expression in males than in females" (271). It also explains that, in the case of gout, symptomatic differences "not only vary from individual to individual; even within the same individual, severity of pain can differ from day to day" (271). The text discusses the ways that internal and external factors can play a role in phenotypic expression of a genotype. It also explains the ways in which monozygotic twin studies can "often help evaluate whether there is a heritable[2] component in the expression of complex traits such as height, weight, diabetes, schizophrenia, and epilepsy" (271). It is useful for the authors to have pointed out the complexities involved in gene expression.

This is especially important because it is done in the context of the seeming simplicity of Mendel's work. It is significant that the authors remind the reader that "Mendel used inbred strains and chose traits that showed a minimal variation under normal environmental conditions." This qualification of Mendel's work has a number of consequences that, unfortunately, are left undeveloped in the text. This qualification serves to contextualize Mendel's work and render it visible as a social construction. It also serves to remind readers that, especially when considering humans, one is talking about much more complex behaviors and genetic interactions. These are all very important qualifications because, if taken seriously, they force teachers and students to avoid oversimplifying the relationships between genes and traits and between genotypes and phenotypes.

Chapter 15 is titled "Mendelian Genetics in the 20th Century." It contains very clear and concise discussions of topics such as "dominance relations," "genetic analysis for allelism," "multiple alleles," "epistasis," the chromosome theory of heredity, crossover and recombination of alleles, gene mapping, and genes and protein synthesis (288–307). The last section, titled "Genes and the Specificity of Proteins," discusses early work on molecular genetics. This discussion foreshadows the topic of Chapter 16, titled "The Molecular Basis of the Gene." Chapter 15 contains one insert titled "Mapping Human Genes" (299). Because the text was published in 1989, it contains no discussion of the Human Genome Project. This insert goes into some detail concerning the ways in which human genes can be mapped and the reasons why this can be useful information. In this regard, it touches on issues of recombinant DNA, genetic engineering, and prenatal diagnosis that will be discussed in more detail in Chapter 18.

Chapter 18 is titled "Genetic Research and Technology." In it the authors state:

> It explores how research in genetics has direct application to humans. Some applications seem highly desirable, others quite questionable. The more complete our knowledge and understanding of genetics, the easier it will be to make these distinctions and to make wise decisions. (355)

The next paragraph begins a discussion of the eugenics movement. The reader is told that

> the eugenics movement began in England during the late 1800s, a period following Darwin's publication of his theory of natural selection as a mechanism of evolution, but before the advent of Mendelian genetics. In the absence of much knowledge of genetics, advocates of eugenics were free to express the socioeconomic prejudices of their time. (355)

The text goes on to cite things such as habitual criminality, feeble-mindedness, alcoholism, and poverty as the kinds of "traits" the eugenics movement sought, in the case of the Nazi regime, to remove from the human genome through sterilization and extermination. The text points out:

> The advent of Mendelian genetics in the early 1900s initially only seems to support these convictions. As the continued absence of any evidence about the hereditary nature of these specific conditions was ignored, stress was placed upon the general importance of heredity. (355–356)

It is interesting to note that here the eugenics movement and its aims and assumptions can be held up to this kind of critical scrutiny, but that something similar cannot be done around the controversy and debate that surrounded sociobiology. Sociobiologists and behavioral geneticists make claims all the time about the inheritance of complex behaviors in humans and nonhumans. They do this often without much support for their position and despite strong critiques from many social scientists, who question their assumptions.

Human sociobiology's exemption from this kind of critical scrutiny in the text is all the more interesting in light of the discussion that follows. The text next raises two questions: "How Different Are the Human Races?" and "What Human Traits Are Inherited?" (356) To answer the first question, the text cites the research that shows that "the genetic variation *among individuals within a "race"* exceeds the variation *between "races"* (356).

When answering the second question the authors say:

> The list of genetic defects numbers in the thousands. Absent from the list are such personality traits as "laziness" or "criminality," which are complex, ill-defined, and heavily influenced by environmental factors. Intelligence is poorly defined by any so-called "IQ" test and is also absent from such a list.
>
> As would be expected from knowledge of how genes function, the listed genetic defects are quite specific. There are some particular inherited metabolic defects associated with a mental handicap, such as phenylketonuria, an infant disease caused by an accumulation of toxins in the body. (356)

This is an excellent reminder to students about the misuse of genetic theory. What is noteworthy here is that the authors are able to present sophisticated arguments that are critical of conceptions of IQ as a "trait," and therefore critical of the idea that that IQ is under genetic control. Yet, they are not able to do the same when considering behaviors such as altruism in the context of discussions of human sociobiology. Perhaps the answer lies in the extent to which a subject has become generally politicized. In the case of IQ, the authors wish to provide a critique of determinist formulations in the debate about the genetics of intelligence, and also wish to highlight the influences of factors such as class, race, ethnicity, and gender.

The text concludes the discussion of eugenics by pointing out that, because most genetic defects are "both rare and recessive," the program was both "unrealistic and ineffective" and "very dangerous when applied to specific subpopulations" (356). This is an important critique for students to hear, because the next section is titled "Genetic Counseling." The text then goes on to describe a variety of abnormal syndromes and the ways to test the genetic makeup of a fetus. The text then details the techniques of recombinant DNA and its applications. Unfortunately, the section ends there and the critique of the early twentieth-century eugenics movement does not carry over into concerns about the possible misuses of new eugenics technologies, such as genetic counseling.

Presentations of Evolutionary Theory

Unit 6 covers the topic of evolution. It is composed of chapters 19 through 22. Chapter 19 is called "Evidence for Organic Evolution" (379). According to the authors, "organic evolution is one of the most interesting, controversial, and emotionally charged issues in science. It is also the single most important unifying concept in biology" (379).

In a section titled "The Modern Synthesis," the authors point out:

> The theory of evolution by natural selection had an immediate and substantial impact on biologists....Nevertheless, the theory was not widely understood; because of its failure to explain inheritance, many scientists questioned the importance of natural selection. . . .
>
> Evolutionary theory has been somewhat modified to what is often called a neo-Darwinian theory, a modification that emerged in the 1930s and gave rise to the so-called "Modern Synthesis."...From this synthesis the great importance of natural selection and *speciation* (Chapter 22) became clear. . . .
>
> The modern synthesis is not essentially different from the theory put forth by Darwin in 1859, except that it incorporates the current understanding of inheritance. It is, however, much more complete because science has gathered a great deal of evidence that was not available to Darwin. . . .
>
> The modern synthesis holds that variation is introduced into a population by random mutations....[S]ome mutations increase fitness of an individual. Usually characteristics that enhance fitness increase in frequency in populations because of natural selection. However, some mutations have little effect on fitness; these mutations may increase or decrease in frequency by chance. . . .
>
> Natural selection is seen as the primary driving force in the origin of families and

> other "higher" groups just as it is in the origin of species. Just as mountains are built up slowly, uplifted by many small steps, a new class of organisms becomes differentiated by the gradual accumulation of small differences, driven primarily by natural selection. (409–409)

This is a very clear and non-ideological statement of what is called the modern synthesis. It would be difficult to justify the claims of Wilsonian sociobiological discourse if this version of the modern synthesis is invoked. So, again, I find it curious that this text contains such contradictions. Perhaps it is the case that the text is trying to reflect the current state of the entire field of biology and, as indicated in Part I of this book, the current state of affairs is controversial and polarized. Also, I find the use of the progressive metaphors in the last section of this quote unfortunate because they convey incorrectly that there is a "progressive" element at work in the process of organic evolution. That is, evolution acts to form "better" or "higher" forms.

In Chapter 21, called "Populations and Natural Selection," the authors make an important point. They say, "The *variation* among individuals in a population is necessary for *natural selection*, and natural selection is an important key to evolution" (413). A little farther down the page, they add: "Evolution is, however, a matter not of discrete changes within individuals but of a cumulative change within a species" (413). It is fitting, in a chapter dealing with population genetics, that the focus should be on populations and not individuals. Yet, while the reader is reminded of this here, it seems to be forgotten in the discussions of sociobiology and animal behavior.

I also find that there are tendencies toward ultra-adaptationist interpretations of the forces acting on evolutionary processes in Chapter 21. In any discussion of natural selection, the understanding of first causes is what keeps the argument from becoming teleological. This is something not always clear in the text. In Chapter 21, the text says:

> An environmental condition may *select for* certain characteristics of some individuals and *select against* those for others. Consider, for example, that some individuals of a population tend to resemble their habitat and thus are camouflaged whereas others may not resemble the habitat so closely and thus are conspicuous (Figure 21.1). If there are predators that hunt by sight, they will see and therefore capture more of the conspicuous individuals. In this way, conspicuous individuals are *selected against* and camouflaged individuals are *selected for*. If there were no variability in a population, there could be no selection. (413)

The text then goes on to describe and show how environmental factors and genetic inheritance produce variability. Unfortunately, there is no mention at that point in the text as to what causes the initial variation. Sexual recombination and various kinds of mutation provide the raw material on which selection works. Starting from these causes and from environmental influences removes teleology from the formulation. Otherwise, in the example quoted above from the text, we would expect to see that all creatures in similar circumstances develop camouflage—because it is being selected for. But this cannot happen unless the "right" mutations or recombinations occur on which selection pressure can act. This point is supposedly what separates the Darwinian and Lamarckian heritages.

When the text does mention the role of mutation later in the chapter, it does so under the general heading "Other Factors Causing Evolution" (425). This is very confusing. The authors do qualify that they are talking about frequencies in a population when they say "there are factors other than natural selection that can cause the *frequency* [author's emphasis] of a characteristic to change in *populations* [author's emphasis] and thus potentially influence evolution. Two such factors are mutation and genetic drift" (425). To be fair, the authors then say:

> Mutations are the ultimate source of all of the genetic variability in populations....However mutations are infrequent, perhaps occurring at each locus in only one in every 10^5 or 10^6 gametes. Thus although mutations are the source of genetic variability they would without selection only slowly cause noticeable changes in allele frequencies in populations. (425)

Yet somehow it seems that the overall effect of the chapter is to weight the emphasis too heavily on the side of selection and adaptation. This is partly accomplished because the role of mutations gets about one-quarter of a page and genetic drift gets about one-half of a page of discussion in the chapter, after about a four and one-half pages discussion of the role of selection.

This chapter does not clarify our understandings of the processes involved in evolution. Rather, it can lead the reader easily into a teleological understanding of evolution or to an ultra-adaptationism. This means that, rather than seeing that selection is limited by the forms thrown up by random mutations and by genetic drift, a teleological formulation weights the emphasis too heavily on the dynamics of selection such that in the teleological formulation it appears that selection causes variation. It makes it appear that selection pressures dictate what can be; instead of understanding that selection pressures form only one set of limits. If you then add to this teleology by ascribing any hint of conscious intention in the process, then what "has evolved" becomes what "should have evolved" instead of what evolved by chance. This is similar to the kind of teleology that infuses Wilson's and Dawkins' talk of selfish genes and talk of organisms as "lumbering robots." It then becomes easy to speak of genes as "causing" behaviors so as to ensure their reproductive success.

In addition to the kinds of problems that I have pointed out above, there are what I would consider to be positive presentations in this chapter. The more that chance and randomness can be built into our understanding of the processes of evolution, the less likely we will be to conceive of evolutionary processes as teleological. Also, I think that the more we acknowledge the complexity that the processes of evolution must entail, then the more accurate conceptions we can develop. The sections I discuss below seem to add that extra bit of randomness and complexity to the discussion.

One of these passages occurs in the brief discussion of mutations, in which the authors talk about the ways in which selection operates on differentially fit characteristics that emerge from mutations.

> If a new characteristic neither increased or decreased fitness, there would be selection neither for nor against it. Such a "neutral" characteristic might increase in frequency simply because of repeated, similar mutations and the lack of selection against it. (425)

Although the text does not take this discussion further, it does at least plant the seed of an idea that the interactions between mutation, characteristics, and selection pressures in the environment are complex and may not proceed in ways that are easily predictable. The nice thing about this example is that it introduces another element of chance into the evolutionary equation.

A second important discussion in the chapter is the one on genetic drift. In this section, the authors skillfully and clearly describe the ways in which the probability of distribution of alleles in a population is calculated. They highlight the role that chance can play in skewing the expected distribution such that the real distributions can be very lopsided, especially in the case of small populations, where certain alleles can be lost because some alleles seem to "'drift' out of the population" (425). Another small but important inset in Chapter 21 is called "Non-Nuclear Inheritance." It deals with cytoplasmic inheritance and, although it is very brief (approximately one-quarter of one page), at least it shows that there is a complexity to inheritance that has yet to be accounted for through the "modern synthesis" (415). And, although the text does not put it in these terms, extra nuclear inheritance opens a door to a model in which environmental influences might somehow become incorporated into the cytoplasmic genetic material and be passed on.[3]

This chapter contains a number of mixed messages in a way similar to the pattern that I saw in *Biology: Discovering Life*. Some issues are covered in ways sympathetic to the human sociobiological enterprise at one point and at other points covered in ways that qualify this support and call the enterprise into question. Also, in the sections on genetics and evolutionary theory, sometimes the presentations are qualified and complex, and at other times they are basic and canonical.

Presentations of the Nature of Science (Knowledge)

This text does not set aside space to discuss the nature of science. There are implicit discussions, however, in the treatments of Mendel, Darwin and more modern theoretical developments. In the case of Mendel, he is treated as a canonical figure and this treatment bolsters a neopositivist orientation. In the other cases, however, the text includes very important contextualizations about their discoveries and about how to conceptualize the influences on the development of science knowledge. In the end, the text backs off from following through to a more clearly "constructivist orientation," and having raised these issues and then waffled, these presentations ultimately may only serve to confuse students or alienate them from pursuing this theoretical orientation.

On the other hand, in Chapter 16 the authors detail the shape and structure of nucleic material and its role in protein synthesis, along with the molecular nature of mutations. In this case, the text does a very balanced job of presenting the discovery of the structure of DNA. It details Chargaff's[4] work on the concentrations of bases in DNA and goes on to say:

> The scene was set for Watson and Crick to establish the physical structure of DNA. In addition to Chargaff's chemical findings, they had access to important physical characterizations of DNA. Rosalind Franklin, working in Maurice Wilkins' laboratory at Cambridge, had X-ray diffraction data that provided critical information on the spatial arrangement of atoms in the DNA molecule. The X-ray data showed DNA contained two (or more) chains wound around each other in a helical manner. While others were working on models with three chains, Watson and Crick in 1953 proposed the *double helix* model. (315)

This is a very different treatment than Mendel received, and it is a well-contextualized, historical presentation in which the reader can see how data from various sources built to the point of Watson's and Crick's model.

Chapter 17, titled "Gene Expression" (333), details the ways in which nucleic acids in prokaryotes and eukaryotes direct protein synthesis. An insert in the chapter discusses the presence of introns in eukaryotic genes. The text presents these noncoding sequences of nucleotides present in DNA, but not within the messenger RNA, as an interesting problem to be addressed. It presents two competing theories based on evolutionary premises. One theory suggests that introns appeared with eukaryotes after prokaryotes had developed and that introns allowed the more sophisticated protein synthesis required by eukaryotes to occur. The other theory suggests that introns have been around as long as prokaryotes, but since prokaryotes do not need them, they are absent in prokaryotes.

The insert concludes with the statement: "it is not possible to conclusively choose between the two models" (345). This is a potentially useful insert, because it shows the reader how little we know even about simple eukaryotes, let alone human genes. In the chapter summary, the authors conclude "that there could be significant differences between prokaryotes and eukaryotes in the regulation of gene expression. The situation in eukaryotes seems more complex and less is known about the details" (351). It also shows the reader how competing theories can be very convincing and that it can be difficult to choose between them. This notion foreshadows work done by Longino, which I discussed in Chapter 3, in which she recommends relying on a modest epistemology when proposing competing explanations for biological phenomena. This discussion could be used to contextualize theory and show students how social factors can affect which theories are given validity, and sometimes hegemony, over others even when both have valid data to support them. This kind of presentation could be developed to show that theory selection is a social process and not a simple assessment of "truth." While the text "plants the seed" of this idea in the mind of the reader, it unfortunately does not develop a systematic account that would allow students and teachers to develop an evaluative critique of determinist claims and science claims in general.

This same chapter delineates a great deal of the evidence for the existence of organic evolution, and at the end of the chapter the authors discuss the status of the theory.

> Although some "theories" are no more than working hypotheses, the theory of evolution is not....[I]t is a useful synthesis of knowledge from many different fields....We do not understand all of the mechanisms that have caused this change; and we do not know and perhaps cannot discover many of the details. The theory of evolution has the same limitations as all theories in natural science: unlike theories in mathemat-

> ics, it cannot be proven without a doubt. It can only be—and has been—demonstrated to be true beyond a reasonable doubt....Organisms have evolved and are evolving today. (395)

It seems obvious that this passage was written to address the popularity of anti-evolutionist views such as creationism. The authors have, in so doing, been forced to enter into the terrain of epistemology in science. As a consequence, this is probably the clearest epistemological statement in the text. It conveys the idea that science does not produce truth but simply produces reliable and accurate descriptions and predictions.

Chapter 20 is titled "The Development of a Theory of Evolution" and, as one would expect, chronicles the historical context of the work of Darwin and Wallace. This is a very good historical chapter, in which Darwin is not canonized. This follows the text's treatment of Crick and Watson that I discussed above. This stance can be clearly seen in the following passage from Chapter 20, which addresses the development of the theory of evolution.

> In many ways, the time was right for such a theory. Selective breeding was enabling farmers to create new vegetable crops and livestock strains. It was increasingly obvious that some species had, in the past, gone extinct. Geologists were arguing that the earth was millions of years old, not merely a few thousand. Perhaps most importantly, it was becoming clear that each different continent or island had its own unique species of plants and animals (a fact, as you will recall from the previous chapter, that is central evidence for the idea of evolution). (398)

Chapter 20 appears to be an unusual chapter in these texts because it is so heavily weighted toward the historical and intellectual climate of Darwin's time. For example, two pages of the chapter are devoted to the influence of Malthus on Darwin's work. The text goes to great lengths to show both what Malthus provides to Darwin and also to show where Malthus was wrong. Unfortunately, the text falls short of attempting to assess whether or not Malthus' contributions make for a stronger or weaker theory—in the long run. There is an assumption that the conception is correct as it stands, and there is also implicit in this conception an assumption that the processes of science will "select" correct interpretations in the long run—no matter where they may have originated.

In another interesting case, *Biology Directions* (Galbraith, 1993)[5], a more recent textbook by the same author, includes an insert on Mendel titled "Mendel's Experiments: Good Science or Good Luck" (Galbraith, 1993, 492). In this insert, the authors bring up the issues that some have raised, and that I discussed in Chapter 4, about Mendel's work. I would say that this is a potentially radical presentation of the process of scientific discovery. In the last paragraph of the insert, however, the authors pull away from this radicalism with the statement: "None of the foregoing is intended to question the integrity of Mendel….Notwithstanding these speculations, Mendel's experimental findings remain a significant contribution to the study of genetics" (Galbraith, 1993, 492). So, in the end, even with this socially more contextualized treatment of Mendel, the ultimate message given to readers is that the strength of the scientific method and the scientific community will triumph over self-interest, power, and social influences.

Summary

In this text, as with *Biology: Rediscovering Life*, I find that the authors do make efforts to create a complex appreciation for the gene/behavior and genotype/phenotype dyads. This text also takes pains to provide at least a glimmer of understanding about the contextual and socially delimited nature of scientific knowledge. This is done, however, in an unsystematic and underdeveloped way that, as I indicated for *Biology: Rediscovering Life*, is likely to prove marginally useful to students and teachers.

Also, this textbook presents what I see as contradictory messages. To be fair, it is also possible that the text appears contradictory not because of the authors' intention to have it appear so, but because this book is a very cooperative effort. So, for example, the sections on genetics may have been formulated by one set of authors while the chapter on animal behavior may have been put together by a different set of authors. Nonetheless, the text presents an uncritical, and I would say hegemonic, view of Mendel, but not of other important figures in the modern history of genetics. It talks about the social context of the development of science knowledge but not in a way that provides readers with any critical tools with which to frame debates around science claims. It talks about the dangers of simplistic models of gene/trait causation in the case of historical eugenics but not for sociobiology or for newly developing biotechnologies. As I indicated above, the contradictory, or seemingly schizophrenic, nature of the presentations cannot lead readers to develop a nuanced sense of the nature of science knowledge. It cannot lead readers to develop a consistent critique of reductive and determinist claims made about the genetic "causes" of human behavior. It is more likely to lead to confusion, to the ignoring of the issues, or to reliance on more dominant reductive and positivist models. In the end, as is likely with readers of *Biology: Rediscovering Life*, I suspect that readers of this text will fall back on these more dominant and reductive conceptual models in which genes have direct and overarching control over traits and behaviors, and on models in which scientific truths emerge ahistorically from the minds of brilliant men.

Notes

1. For a more thorough description of this, see Chapter 4.
2. See Note 2 in Chapter 5.
3. Jan Sapp's work on the history of genetics detailed how research on cytoplasmic inheritance lost out as genetics, and later molecular genetics, gained dominance in research in genetics and evolutionary theory (Sapp, 1987; 1990). See also E. Keller, 1995.
4. Chargaff's work in the post–World War II years indicated that the four nucleotide base pairs present in DNA were not present in equal proportions and that the nucleotide composition varied in complex ways. Most importantly, he discovered that in all cases, the amount of adenine is equal to the amount of thymine, and the amount of guanine is equal to the amount of cytosine.
5. This textbook is more recent. It is designed for lower-level courses and has no chapter on animal behavior.

· 8 ·

TWO EDITIONS OF *BIOLOGY*

Moving Toward More Balanced and Nuanced Positions

Part I: The Fourth Edition

As I indicated in Chapter 4, I have used the fourth edition of *Biology* by Neil Campbell (1996) even though Circular 14 included only the third edition (1993). The updating of Circular 14 ended before the publication of the fourth edition. For a more effective comparison, however, I am using the fourth edition because its publication timeframe is similar to the other books that I have reviewed. In addition, this text is unique in that the author includes interviews with prominent biologists who specialize in fields covered by the units of the text, and in this edition E. O. Wilson is one of the biologists interviewed. I was interested in seeing if the selection of Wilson would be reflected in any way in the textbook's treatment of sociobiology.

Presentations of Animal Behavior and Sociobiology in the Fourth Edition

The last chapter of the textbook, Chapter 50, is titled "Behavior." In this regard, it is like all the other texts in dealing with the topic in one chapter at the end of the book. Campbell does not shy away from linking evolutionary theory and behavior and early in the chapter defines "behavioral ecology" as "the research approach based on the expectation that animals increase their Darwinian fitness by optimal behavior" (1173). He goes on to give a very thorough account of the elements of theory and research that compose the field of animal behavior studies. He also provides some historical context by explaining to the reader that "before the advent of behavioral ecology, an older discipline called *ethology* dominated the study of animal behavior" (1176). He then provides a thorough overview of the findings in ethology.

Campbell then goes on, in the next section, titled "Learning Is Experience-Based Modification of Behavior," to discuss the nature-versus-nurture controversy as it originally played out between European ethologists and North American behavioral psychologists. Again, he has included important context for the debate, thereby providing students with a way to frame the debate itself. He ultimately takes a clear position in the middle ground on the nature-versus-nurture controversy.

> Few scientists ever believed that behavior is either all genetic or all learned. Instead, the debate was mostly about which input—instinct or learning—is of primary importance.
>
> Nearly all biologists today agree that most behavior is a consequence of genetic and environmental influences. . . .
>
> Despite the dictum that behavior is some mix of learning and instinct, or environment and genetic inputs, it is useful to sort out the relative contributions of genetic and environmental influences on a particular behavior. (1180)

Over the next ten pages, he then provides examples of both scenarios in which either environmental or genetic factors seem to be most influential for particular animal behaviors. Again, his even-handedness in covering the topic of behavior and in tackling the controversy about the relative influences on behaviors allows readers to develop informed opinions of their own while, at the same time, providing clear direction and support for the process of scientific investigation that serves as an epistemological anchor for all who wade in these treacherous theoretical currents. He provides a way to deal with controversy in science, and at the same time he suggests a faith in the method of investigation.

Campbell introduces the concept of sociobiology about halfway through the chapter, in a section titled "Sociobiology Places Social Behavior in an Evolutionary Context" (1191). He introduces sociobiology as a relatively new field, employing "evolutionary theory as the foundation for the study and interpretation of social behavior" (1191). He also includes a reference to the interview with Wilson that is included in the textbook at the beginning of Unit 5. So, while he has included in the textbook many issues that are implicitly critical of Wilson's work, he has included not just an interview with Wilson but also a discussion of sociobiology. Over the next eight pages, he covers topics such as agonistic behavior, dominance, territoriality, mating, and modes of communication but does not specifically relate these to sociobiological principles. Then, in a section titled "The Concept of Inclusive Fitness Can Account for Most Altruistic Behavior" (1200), he introduces and explains the concepts of altruism, inclusive fitness, and kin selection to help explain how altruistic behavior can be seen as adaptive behavior. In this section, he refers to W. D. Hamilton and J. B. S. Haldane as developers of these theories but does not refer to either Wilson or sociobiology in the discussion.

In the next and final section of both the chapter and the book, titled "Human Sociobiology Connects Biology to the Humanities and Social Sciences" (1202), Campbell devotes the most space to sociobiology and, in this case, specifically to human sociobiology. In so doing, he explains that Wilson's work

> speculates about the evolutionary basis of certain kinds of social behavior in humans....[It also] rekindled the nature-versus-nurture controversy, and the debate over sociobiology remains heated two decades after its publication. (1202)

He then uses the example of incest to begin a discussion of the relative influences of culture and genetics on complex human behaviors. He concludes with a number of strong statements on the controversy.

> Some sociobiologists, including Wilson, envision cultural and genetic components of social behavior as linked in a cycle of reinforcement. . . .
>
> The spectrum of possible social behaviors may be circumscribed by our genetic potential, but this is very different than saying that genes are rigid determinants of behavior. This is the core of the debate about sociobiology. (1202–3)

Campbell then goes on to explain that most sociobiologists do hold that both culture and genetics influence human behavior. He then ends the section as follows:

> Perhaps it is our social and cultural institutions that make us truly unique and that provide the only feature in which there is no continuum between humans and other animals. (1203)

This seems to make the case that we cannot simplistically apply the principles used in animal sociobiology to human behaviors and societies. The last figure in the book is of an elder teaching a young person a traditional dance. The caption for figure 50.24 reads:

> Both Genes and Culture Build Human Nature. Teaching of the younger generation by the older is one of the basic ways in which all cultures are transmitted. Sociobiologists see tutoring as an innate tendency with adaptive value that has evolved in the human species.

The presentation of sociobiology and of the nature/nurture controversy in general is fundamentally different in this text from the positions in the other texts of this generation that I examined. In the other texts, one saw a lot of "mentioning" of environmental influences, but little substance. At best, those presentations appear, to me, to lead only to confusion. In this case, however, Campbell has taken pains in many areas of the text to ensure a prominent place for environmental influences on phenotypes and ultimately on behavior.

Presentations of Genetics in the Fourth Edition

There are eight chapters (12 through 19) that make up the "Gene" unit in this textbook. In Chapter 13, Campbell tackles head-on the thorny issues of both simple Mendelianism and the nature/nurture debate. First, Campbell provides one of the few clear definitions of character and trait provided by any of the textbooks examined. According to Campbell, "geneticists use the term *character* for a heritable feature, such as flower color, that varies among individuals. Each variant for a character, such as purple or white flowers, is called a *trait*" (239). Then he speaks of Mendel's work and notes:

> It was brilliant (or lucky) that Mendel chose pea plant characters that turned out to have a relatively simple genetic basis....The relationship between phenotype and genotype is rarely so simple.

This is an important qualification, because it indicates to the reader that this text is not focusing on or advantaging a simple Mendelian model nor on a simplistic genetic reductionism.

Ultimately, there is a section in this chapter titled "Nature Versus Nurture: The Environmental Impact on Phenotype" (251). In this section, the author makes a specific effort to point out to the reader that "phenotype depends on environment as well as on genes" (251), and that even identical twins develop phenotypic differences based on their different environmental exposures. He sums up this idea in the following quote:

> Whether it is genes or the environment—nature or nurture—that most influences human characteristics is a very old and hotly contested debate that we will not attempt to settle here. We can say, however, that the product of a genotype is generally not a rigidly defined phenotype, but a range of phenotypic possibilities over which there may be variation due to environmental influence. (251)

As part of this effort at highlighting the ways in which non-Mendelian inheritance occurs, in the last section of Chapter 14, titled "Extranuclear Genes Exhibit a Non-Mendelian Pattern of Inheritance" (278), Campbell points out that both mitochondria and plastids—which are located in the cytoplasm of cells—"replicate and transmit their genes to daughter organelles" (278). He also points out that "these cytoplasmic genes do not display Mendelian inheritance" patterns.[1] All in all, Campbell, in this unit, provides both an excellent and comprehensive discussion of the state of genetic theory and a way for the reader to think about the complexities of the interplay between genotype and phenotype, while pointing out to the reader that the heritability of traits is more complex than a simple Mendelianism would dictate. It would appear that the author of this textbook is aware of the larger debates initiated by the sociobiology debate and he attempts to address some of these issues in an indirect or embedded fashion.

Presentations of Evolutionary Theory in the Fourth Edition

As with all these texts, a great deal of space is devoted to evolutionary theory. In this text, it is broken down into two units. The first is Unit 4, titled "Mechanisms of Evolution." It includes four chapters: 20 through 23. The second unit, Unit 5, is titled "The Evolutionary History of Biological Diversity." It includes seven chapters: 24 through 30. Chapter 21 is titled "The Evolution of Populations." This is an interesting chapter not because it takes aim directly at an ultra-adaptationist position, but because of its focus on populations as the units of evolution. In this chapter, the author introduces the idea of the modern synthesis in biology and the importance of populations as the units of selection within this synthesis. At the same time, he also qualifies the permanency of this paradigm and any scientific paradigm by instructing the reader that the modern synthesis is also undergoing challenges as new theories emerge, and that this is and should be a normal part of the scientific process (417). For its time, this text covered material that most other texts did not. Campbell does three important things in this section. He again shows scientific practice and theory as process, he introduces the idea that populations are key units of evolution, and he acknowledges that there are new theoretical positions that are challenging this established paradigm. This last step also acknowledges some of the new developments that have emerged from within the sociobiological discourse, such as work that looks at genes as a locus of evolution instead of populations. This kind of contextualization is exactly what science students should be getting, and it is these kinds of ideas that will empower students and scientists as they make sense of debates and controversies such as the sociobiology debate.

This textbook takes an indirect position on the question of ultra-adaptationism. First, in Chapter 21, in a section titled "Does All Genetic Variation Affect Survival and Reproductive Success?" he points out: "Some of the genetic variations observed in populations are probably trivial in their impact on reproductive success" (429). He cites the example of human fingerprints and also some of the variations around human hemoglobin as examples of the "theory of neutral evolution." This is similar to the kinds of arguments that have been made by Gould, an ardent critic of sociobiological discourse (see Chapter 1 in Part 1 of this book). Campbell includes this position in the chapter in a neutral manner, and he also qualifies this theory by explaining that it is unclear how much, if any, variation is really neutral, or whether it is really a question of our not being able to see the connections to conferred selective advantage. Again, just the fact that he has introduced the reader to the possibility of neutral selection is an important step, because it allows the reader to consider formulations that are not ultra-adaptationist.

In this same chapter, he also goes on to link a qualification of simple Mendelianism with a qualified adaptationism and a qualification of the idea that genes are important units of selection. In a section titled "What Does Selection Act On?" he warns the reader that selection is acting on the phenotypic presentation of an organism rather than directly on the genotype itself. This is especially important, because it may take many gene loci to control one phenotypic trait. He gives the example of human height and explains that with respect to this trait:

> Individuals in a population do not usually fit into exclusive categories...but instead vary continuously over a phenotypic range [and such] quantitative characters usually account for most of the variation exposed to natural selection. (430–431)

He goes on to say:

> The finished organism subjected to natural selection is an integrated composite of its many phenotypic features, not a collage of individual parts. (431)

These debates about the importance of genes, organisms, and populations in the selection process have been important within the sociobiology debate. Campbell presents some of the issues in this chapter without taking a direct position on the debate. In so doing, he does not present one side as more legitimate than the other and does acknowledge that science practice, as a process, will change paradigms as time and knowledge change. All in all, he is unique among the textbook authors that I have looked at in that he takes up positions on these topics in the textbook's units on evolutionary theory.

Presentations of the Nature of Science (Knowledge) in the Fourth Edition

In some respects and for its time, this fourth edition textbook has a sophisticated introductory chapter. While it does not try to contextualize the process of scientific inquiry in ways that the more recent textbooks that I examine do, it does have a very important discussion on "emergent properties" in biological systems that is similar to the discussion in the eighth edition of *Biology*. that I discussed in Chapter 5. In the introductory chapter, "Introduction: Themes in the Study of Life," the author presents a number of ideas that highlight the primacy of studying organisms and not just components of organisms. This is an important idea, because it countermands the sociobiological focus on using reductionism as a bona fide method to understand the behavior of organisms. This is one of the themes that Campbell discusses in the chapter titled "Each Level of Biological Organization Has Emergent Properties." In this section, the author tells the reader:

> With each step upward in the hierarchy of biological order, novel properties emerge that were not present at the simpler levels of organization. These emergent properties result from interactions between components...and an organism is a living whole greater than the sum of its parts. (4)

And further on in this same chapter, when discussing the theme titled "A Feeling for Organisms Enriches the Study of Life," the author points out:

> While cells are the units of organisms, it is organisms that are the units of life. It is an important distinction....We may go on to study life at different levels of organization,

> but it is organisms we must keep in mind if our exploration of cells or ecosystems is to be meaningful. (8)

The author goes on in that section to talk about Barbara McClintock's work on corn, and the importance of her deep emotional commitment and knowledge of the plants with which she worked. Campbell points out, based on Keller's (1983) biography of McClintock: "It is a common misconception that scientific objectivity demands that biologists be detached from the organism they study" (9). Rather, citing Keller, he points out that good science requires "deep emotional" commitment. Campbell also points out in this chapter, through two other themes, that "Evolution Is a Core Theme of Biology" and that "Science As a Process of Inquiry" involves "hypothetical-deductive thinking" (13–20). This latter theme is used to highlight a very good discussion of the ways in which research is conducted and hypotheses formed and developed. Thus, even though this chapter does not contain a sophisticated epistemological discussion on the nature of science and science knowledge, it does contain a discussion of organisms as important biological actors. In this way, the author both anticipates and shadows this same point that has been made by numerous critics of human sociobiological discourse and thereby provides the reader, indirectly, with important conceptual information with which to judge the claims of sociobiological discourse and an important issue that has arisen through the sociobiology controversy.

With respect to canonical figures such as Mendel and Darwin, Campbell's presentations are not as sophisticated as Galbraith's *Biology Directions* (1993). In Chapter 13, however, titled "Mendel and the Gene Idea," Campbell does point out that, because Mendel brought "an experimental and quantitative approach to genetics," he was successful and that the example of Mendel again illustrates "science as a process" (238). This is an interesting reference, because Campbell does not try to contextualize Mendel or his work in a way that supports a social reading of his discoveries. Nor does he attempt to provide any historical deconstruction of Mendel's work as was done by Galbraith. This latter phrase, however—"science as a process"—had begun to be used by philosophers, such as Hull (1988), who were exploring the nature of science in the late twentieth century. This allusion shows that Campbell is aware of the work in science studies being done on the nature of science.

In addition, Chapter 20 focuses on Darwin, and, while he is given canonical status in this chapter, the chapter does this while providing the historical context for the emergence of the theory of evolution. Figure 20.1 (400), titled

"Darwin in Historical Context," graphically locates Darwin within a two-hundred-year timeframe, and shows the timeframes for relevant theorists and theories that came before and after Darwin. In that sense, it is an excellent introduction to some of the important ideas and debates that ranged before and during Darwin's life. Even though Campbell does nothing overtly to challenge a traditional understanding of the nature of science, he does provide important historical context for Darwin's time and his theoretical development.

Summary of the Fourth Edition

All in all, this is a very successful textbook, both as a fine introduction to the vast topic of biology and also as an introduction to some of the important issues that have emerged as a result of the influence of sociobiology and the sociobiology debate. It stays away from simple Mendelian explanations and ultra-adaptationist formulations of evolutionary theory. At the same time, it provides good coverage of the issues surrounding the nature/nurture controversy while doing more than just "mentioning" environmental influences. Instead, these issues are given a strong place in his conceptual framework. In addition, of the texts I reviewed that were published in this decade, this text and its later edition provide the only real attempts at defining the terms "trait" and "character." And, even though the material on the nature of science does not contain a sophisticated epistemological discussion on the constructionist nature of science and science knowledge, it does contain a discussion of organisms as important biological actors. In this way, Campbell shadows these same points that have been made by numerous critics of Wilsonian sociobiological discourse. He thereby provides the reader indirectly with important conceptual information with which to judge the claims of that discourse as it relates to the issue of the conception of organisms. I would argue that Campbell has provided readers with much useful information that can allow them to develop informed opinions about the sociobiology debate.

Part II: The Seventh Edition

In this seventh edition of *Biology*, Jane B. Reece is added as a co-author with Neil Campbell (2005); the publisher is Pearson Educational (as Benjamin Cummings). This text has enhanced most of the arguments put forward in the fourth edition text in all the areas that I examine. The idea of science as a social process has not been highlighted as it was in the fourth edition. The idea of systems biology is more developed in this edition, and I discuss it in the nature of science section. A new idea expressed by the authors of the seventh edition is an awareness of the constructivism debate in science education. The authors display this awareness in a preface subsection titled "Balancing Inquiry with a Conceptual Framework" (ix), in which they say that, while they want to emphasize the importance of "practice" for students in the development of science knowledge,

> *Biology*, Seventh Edition, is *not* a "reform textbook" of the genre that replaces a careful unfolding of conceptual content with a stream of relatively unconnected research examples, requiring beginning students to put it all together for themselves.

The authors here signal their awareness of constructivist theory in science education and their position on the controversy surrounding such "reform textbooks." In this sense, they attempt to provide a kind of middle ground position that acknowledges science as a social process and at the same time maintains a place for science practice as a special or unique generator of knowledge. The authors believe that one of the best ways for the text to support student inquiry is "by providing context with clear, accurate explanation of the key biological concepts" (ix).

Presentations of Animal Behavior and Sociobiology in the Seventh Edition

The chapter on behavior is called "Behavioral Ecology." This is an acknowledgement of the post-sociobiological "evolution" of the study of biology and behavior that I discussed in the Introduction and in Chapter 1 of this book. These authors employ the concept in ways that I would consider to be positive in that, in the hands of these authors, this concept frames the chapter as the study of the interactions within an ecological system and not simply an unfolding of the interplay between genes and natural selection. These latter elements are not missing from the chapter but are interwoven into the notion of a complex interplay of forces at work. A number of key concepts are highlighted in the chapter. The first concept, 51.1, titled "Behavioral Ecologists Distinguish Between Proximate and Ultimate Causes of Behavior," divides the study of behavior into the immediate stimuli and mechanisms that trigger a behavior as juxtaposed to the evolutionary significance of that behavior. Proximate causes are considered the "how" questions, and ultimate causes are considered the "why" questions (1107). This distinction is not unique to this text but is rather a "normal" way to separate these dimensions. Here, however, in the overall context of the way this chapter is framed, it feels less like lip service than an honest attempt to signal the reader that environmental issues are a key component of behavior and that behavior arises as an interaction of the two realms. In this regard, they begin the discussion of concept 51.2, titled "Many Behaviors Have a Strong Genetic Component," with the statement:

> Extensive research shows that behavioral traits, like anatomical and physiological aspects of a phenotype, are the result of complex interactions between genetics and environmental factors. The conclusion contrasts sharply with the popular conception that behavior is due either to genes (nature) or to environment (nurture). In biology, nature versus nurture is not a debate. Rather, biologists study how *both* genes *and* the environment influence the development of phenotypes, including behavioral phenotypes. (1109)

This is a very clear statement on their overall position. They see both as important factors, and they illustrate this through their inserted examples on selected animal behaviors. In these cases, both proximate and ultimate causes are included in the examples given. This is not a form of "mentioning"; both sides are given substantial weight. In general, this is a very well-written chapter, rich in detail and including complex discussions of the genetic, evolutionary, and environmental influences on behavior. It is also interesting that the insert titled 51.6 includes concepts of inclusive fitness, kin selection, and reciprocal altruism, but, in a fashion similar to the fourth edition, they are discussed only as contributions made by William Hamilton, especially as they relate to the study of altruistic behavior in animals (1128–1132).

The final subsection in the chapter is titled "Evolution and Human Culture" (1132–1133). The authors make clear their position on the importance of culture in the subsection that precedes "Evolution and Human Culture," titled "Social Learning" (1130–1131). This subsection opens with the following:

> When we discussed learning earlier in this chapter, we focused largely on the genetic and environmental influences that lead animals to acquire new behaviors. Learning can also have a significant social component, as seen in *social learning*, which is learning through observing others....Social learning forms the roots of culture....(1131)

The final subsection of the chapter and the text itself ends with an inset, Figure 51.38. This is a photo of an older man and a young boy working with fishing tackle and the caption reads:

> *Both genes and culture build human nature.*
> Teaching of a younger generation by an older generation is one of the basic ways in which all cultures are transmitted. (1132)

The authors make clear in this subsection that it is not just environmental factors but "social" factors as well that are crucial to "building" and not discovering human nature. And, in addition to a different photo, the caption differs from the fourth edition in that sociobiology is not discussed or related to this discussion of genes and culture.

One significant development in this seventh edition is that no mention is made of Wilson's synthetic text and conceptual framework at all in the chapter, and the word "sociobiology" is not mentioned until the last page of the chapter in the final subsection—and then it is only discussed in terms of human behavior. This is somewhat unusual, as one could argue that they present concepts important to all sociobiology in this chapter. The authors appear to feel, however, that the word itself is only relevant to "mention" in terms of human behavior.

At the same time, the authors relegate sociobiology to a discussion of the controversial debate that surrounds conceptualization of the determinates of human behaviors. This in itself is interesting, because so little of Wilson's main text was devoted to discussions of human behavior. In this section on "Evolution and Human Culture," the authors present the positions in the debate concerning the influences of genes, environment, and culture on human behaviors. This subsection is half a page in length. They present the issues as forming a controversy with Wilson and sociobiology standing for the premise that "certain behavioral characteristics exist because they are expressions of genes that have been perpetuated by natural selection" (1132). They also acknowledge the recent discoveries that link genes to "complex behavioral traits, such as depression, violence, or alcoholism" (1133). This is one of the few textbooks that I have reviewed that presents and discusses the general controversy created by Wilson's text and by the discourse around sociobiology that developed from it. They then go on to quote Robert Plomin, director of the Center for Developmental and Health Genetics at Pennsylvania State University, who believes that "research into the heritability of behavior is the best demonstration of the importance of the environment" and that genetic, nongenetic, and environmental factors build and influence each other (1133).

Ultimately, the subsection and the chapter end with the sentence:

> Perhaps it is our social and cultural institutions that make us truly unique and that provide the only feature in which there is no continuum between humans and other animals. (1133)

The fourth edition also ended in a similar manner and with the same overall message about the importance of environmental, social, and cultural influences on human behavior. What has been done in this edition is to present the issue as a controversy or debate and to give space to both sides.

Presentations of Genetics in the Seventh Edition

Sections 13 through 21 of this text are devoted to genetics. In the years since the fourth edition, this section has been enhanced to reflect developments in genetics and proteinomics. These chapters provide an extensive and in-depth coverage of the topic. Chapter 14, "Mendel and the Gene Idea," is the most relevant for this study. In this chapter, the focus is on Mendel as a key figure in the development of genetics who used the methods of science to make his discoveries. In a fashion similar to the fourth edition and early in the chapter, the authors define the concepts of character and trait. This is important because most introductory biology textbooks do not provide definitions, and if they do, the definitions are usually circular. These authors not only provide definitions but also note the ambiguity present in them.

> A *character* is a heritable feature, such as flower color, that varies among individuals. Each variant for a character is called a *trait*. (Some geneticists use the terms *character* and *trait* synonymously but in this book we distinguish between them.) (252)

They go on to present Mendel's laws of segregation and independent assortment in the following subsections. These sections are clear and well-presented descriptions of Mendel's experiments.

There is a very important discussion in the section titled "Concept 14.3: Inheritance Patterns Are Often More Complex Than Predicted by Simple Mendelian Genetics" (260). In this section, the authors say that modern genetics has revealed inheritance patterns that are more complex than Mendel imagined and that he was lucky to have chosen the pea plants he did because they had a "relatively simple genetic basis" (260). They also go on to say, "The relationship between genotype and phenotype is rarely so simple." So, while this is not as nuanced a treatment of Mendel as Galbraith's *Biology Directions* (1993), it is important that such a canonical figure is given both context and limitations. It is also important vis-à-vis simple Mendelian presentations used in determinist arguments that the authors tell students that these simple presentations are rarely the whole story.

The authors reiterate this idea in a later section in the chapter titled "Nature and Nurture: The Environmental Impact on Phenotype." In this section, the authors discuss the relationship between polygenic characters such as skin color and the environment.

> Environment contributes to the quantitative nature of these characters as we have seen in the continuous variation in skin color. Geneticists refer to such characters as *multifactorial*, meaning that many factors, both genetic and environmental, collectively influence phenotype. (264)

The point is reiterated again in the next section titled "Integrating a Mendelian View of Heredity and Variation." In this section, they state clearly:

> Over the past several pages, we have broadened our view of Mendelian inheritance by exploring the spectrum of dominance as well as multiple alleles, pleiotropy, epistasis, plygenic inheritance, and the impact of the environment. How can we integrate these refinements into a comprehensive theory of Mendelian genetics? *The key is to make the transition from the reductionist emphasis on single genes and phenotypic characters to the emergent properties of the organism as a whole, one of the themes of this book* [author's emphasis]. (264)

This is the clearest anti-reductionist statement that I have found in any of the textbooks that I have examined, and it constitutes a substantive development from the presentation in the fourth edition. It clearly speaks against reductionism supported by a simple Mendelianism and in favor of a systems biology approach that privileges organisms over individual genes. Again, in Chapter 15, titled "The Chromosomal Basis of Inheritance," the authors introduce Concept 15.5, titled "Some Inheritance Patterns Are Exemptions to the Standard Chromosomal Theory" (288). In this context, they present discussions of genomic imprinting and the inheritance of organelle genes. Also, in Chapter 19, the authors introduce the concept of "epigenetic inheritance," defined as "the inheritance of traits transmitted by mechanisms not directly involving the nucleotide sequence" (364). And they go on to say: "Researchers are amassing more and more evidence for the importance of epigenetic information in regulation of gene expression" (364). All of this work adds to the picture of complex and varied interactions that connect genetic material to phenotypes and to behaviors as opposed to presentations of these interactions through the lens of simple Mendelianism.

Presentations of Evolutionary Theory in the Seventh Edition

This seventh edition devotes a lot of space to evolution. It has two units devoted to evolutionary theory. Unit 4 is called "Mechanisms of Evolution" and it contains four chapters. Unit 5 is called "The Evolutionary History of Biological Diversity" and it contains nine chapters. Chapter 22, the first chapter in Unit 4, is called "Descent with Modification: A Darwinian View of Life." In this same chapter, in the context of a discussion of the importance of skepticism in the development of scientific theories, the authors use the example of the role of natural selection in evolution to make the point.

> The skepticism of scientists as they continue to test theories prevents these ideas from becoming dogma. For example, many evolutionary biologists now question whether natural selection is the only evolutionary mechanism responsible for the evolutionary history inferred from the fossil and molecular data. As we'll explore further in this unit, other factors have also played an important role, particularly in the evolution of genes and proteins. (451)

In so doing, the authors provide not just excellent and comprehensive discussions of the dynamics that operate in the processes of evolution, they also undermine the ultra-adaptationist focus that is so crucial to determinist arguments.

This textbook is also one of the only texts to indicate clearly that "evolution is not goal oriented" (486) and also "why natural selection cannot fashion perfect organisms" (469). In the best senses, evolution is treated not dogmatically or reverently but as a viable and powerful scientific theory subject to influence, context, and revision in the best traditions of science. This is in keeping with the notion of science knowledge as an ongoing project. And, in keeping with this theme, Chapter 25, titled "The Origin of Species," is an excellent treatment of the current state of information on how species evolve and differentiate. Most importantly, this chapter also introduces the concepts of anagenesis and cladogenesis (472) and describes the two basic processes that operate at the level of speciation. Much of the data that makes this differentiation possible is derived from recent studies at the molecular genetics level. In subsequent chapters in Unit 5, the authors spend considerable effort to show the reader how these new data have begun to force us to "revise our understanding" of the ways we have categorized all creatures in the tree of life (529). All of these examples also help to demonstrate the "process of science" that the authors talk about early in the text.

Presentations on the Nature of Science (Knowledge) in the Seventh Edition

There are big developments in this edition over the fourth edition in this text's treatment of the nature of science. These developments are in the discussions of the limits of reductionism as a method of scientific analysis, in the discussions of the importance of a "systems" approach in biology, and in the discussion of the limitations of scientific inquiry. Chapter 1 deals directly with issues concerning the nature of science and opens with a listing of key concepts to be covered in the chapter. The subsections covered by Concept 1.1, titled "Biologists Explore Life from the Microscopic to the Global Scale," talk about the hierarchy of biological organization from the biosphere level down to the molecular level. In these subsections, the ideas of heritability and DNA are introduced as well. This idea of levels of analysis will be important throughout the text as the authors emphasize a systems approach to the study of biology.

In a section in the preface titled "New to the Seventh Edition" (xxii), we are told that new to the first chapter and to the book is a "discussion of systems biology as one of the book's themes" and that the section on scientific inquiry is "more robust." The addition of systems biology also speaks to the attempt to present more of the complexity that is a key part of living systems. The subsections in the first chapter covered by Concept 1.2, titled "Biological systems are much more than the sum of their parts," are new to this edition and are interesting in their look at the complexity of life systems. The authors discuss the emergent properties of systems as complexity increases. They also discuss what are called "The Power and Limitations of Reductionism." This is the heading of the subsection and in it the authors point out both the strength in learning about things by taking them apart and also the reality of the way this process disrupts the integrity of organisms.

> . . . [W]e cannot fully explain a higher level of order by breaking it down into its parts. A dissected animal no longer functions; a cell reduced to its chemical ingredients is no longer a cell. Disrupting a living system interferes with the meaningful understanding of its processes. (9)

We can see here that the authors attempt an epistemological middle ground.

In part, the authors explain that the deluge of bioinformatics generated by the Human Genome Project and its offshoots are forcing us to take a systems approach to make sense of the information (10–11). They also argue that a multidisciplinary approach is required. In this section, they also present the idea of "feedback regulation in biology." In the rest of the chapter, as they cover the remaining core concepts, they touch on the organization of life forms, evolution, and forms of inquiry. All of this presentation is important because it undermines the value of using a simple reductionism to understand complex biological systems. In so doing, this presentation undermines the simple reductionism that is one of the epistemological foundations of Wilsonian sociobiological discourse.

Another interesting and innovative element in this text is the very conscious and deliberate way in which the authors explain what they term "the limitations and the culture of science" (24–25). In the preface, they include a subsection titled "Modeling Inquiry by Example," (vii) in which the authors point out:

> Scientific inquiry has always been one of *Biology*'s unifying themes. Each edition has traced the history of many research questions and scientific debate to help students appreciate not just "what we know" but "how we know," and "what we do not yet know."

The subsections in Chapter 1 that are organized around Concept 1.5, titled "Biologists Use Various Forms of Inquiry to Explore Life," focus on forms of inquiry, and there are some very good presentations that describe inductive and deductive research processes. In the subsection titled "The Myth of the Scientific Method," the authors explain:

> Very few scientific inquiries adhere rigidly to the sequence of steps prescribed by the "textbook" scientific method…[and] discovery science has contributed much to our understanding of nature without most of the steps of the so-called scientific method. (21)

Also, in the subsection titled "The Culture of Science," the authors make two important points. First, they point out that, as more women have entered the field of biology, this has "affected the emphasis in certain research fields" (25). They cite the example that when mating habits of animals were studied primarily by male biologists they were more likely to focus on the competitive behaviors of males, and that as more women have entered the field there is now more study of the role of females in choosing mates. Further on in this same subsection, the authors signal their awareness of debates in the philosophy of science around social and cultural influences on science knowledge.

> Some philosophers of science argue that scientists are so influenced by cultural and political values that science is no more objective than other ways of "knowing nature." At the other extreme are people who speak of scientific theories as though they were natural laws instead of human interpretations of nature. The reality of science is probably somewhere in between—rarely perfectly objective, but continuously vetted through the expectation that observations and experiments be repeatable and hypotheses be testable and falsifiable. (25)

This passage, taken together with the authors' other qualifications about reductionism, appears to indicate that they are aware of some of the recent debates around the nature of science that I covered in Chapter 3. They are clearly choosing to present a more modest set of claims about science practice and science knowledge in which social influences have significance.

To reinforce this idea of the social dimension in science, they include in this chapter a subsection on "Science, Technology, and Society,"' and, in this subsection, the authors point out the interconnections between science, technological adaptations, and effects on society (25–26). Also, in Chapter 22, the first chapter in Unit 4, called "Descent with Modification: A Darwinian View of Life," they do a very good job of presenting the historical context for Darwin's theories. This contextualization helps to take Darwin and the process of scientific discovery off a pedestal and highlights the importance of social setting. According to the authors:

> The impact of an intellectual revolution such as Darwinism depends on timing as well as logic. To understand why Darwin's ideas were revolutionary, we need to examine his views in the context of other Western ideas about earth and its life. (438)

All of this contextualizing of the work of scientists and the limitations put on the claims of science together form a much more sophisticated presentation of the nature of science than any of the other textbooks I analyzed.

Summary of the Analysis of the Seventh Edition

The seventh edition exten ds the ideas that were presented in the fourth edition in most areas. The presentation of sociobiology has shifted such that discussion of sociobiology itself has almost vanished from the behavior chapter. The presentations of genetics and evolutionary theory are more nuanced, and effectively make simple Mendelian and ultra-adaptationist interpretations much more difficult to support. In this regard, this text has the clearest anti-reductionist statements concerning the relationship between biology and human behavior of any of the texts I examined. In addition, when defining the terms "character" and "trait" the authors indicate that many geneticists conflate the two terms, thus indicating that these terms are a site of confusion in the discipline. Probably the biggest advance in this edition over the fourth edition, and over all the other texts that I analyzed, is in how this edition presents the issues surrounding the nature of science. Clearly the authors are making the attempt in the new edition to contextualize the process of inquiry for students and to animate student learning.

Note

1. This discussion echoed Sapp's (1987) work on the history of the struggles around the acceptance of cytoplasmic inheritance in a fashion similar to Galbraith's discussion of Mendel's work that echoed work done by other historians of biology.

· 9 ·

WHERE TO GO FROM HERE?

Part I of this analysis was an examination of the history of the sociobiology debate. Although there were many cogent critiques, especially of various aspects of E. O. Wilson's work, at the time these critiques did not seem to slow down or discourage people from extending this paradigm and did not seem to slow down or discourage funders and publishers from nurturing its development. I have presented a number of factors that have encouraged the rapid expansion of nonhuman and human sociobiological discourse and related disciplines. This has included a discussion of the agendas and needs of publishers and the ways in which sociobiology and strong genetic and biological determinisms in general resonate with larger dominant social discourses.

In addition to these real and external forces, this analysis shows that it is crucial to understand and "deconstruct" the ways in which sociobiology constructed its legitimacy. Wilsonian sociobiological discourse derived much of its legitimacy from its selective interpretations of canonical genetic and evolutionary theory and its juxtaposition of itself as a natural and therefore "true" science in opposition to the softer, inaccurate social sciences. Also, this discourse enhanced its legitimacy by constructing very hegemonizing and delimiting presentations—which, in effect, limited the boundaries of the debate. This delimiting functions so as to include specific interpretations of genetic theory, evolutionary theory, and a conception of the nature of science. Taken together with principles such as kin selection and reciprocal altruism, this discourse sought to minimize social influences in ways that reduced organisms to mere biochemical entities. In this process, the cognitive, social, and cultural dimensions of these organisms (both human and nonhuman) and, in effect, the organisms themselves, were reduced to genetic and biological processes. Ultimately, critics and proponents of the general human sociobiological project have operated with different conceptions of what organisms are and what a science of these organisms should look like.

Wilson's work was, in part, an attempt to correct what he saw as unbridled and indefensible extreme environmentalist explanations of human behavior that took no account of biological factors and constraints. He wanted to introduce biology, and especially genetic and evolutionary theory, back into the equation. Wilson went about creating *Sociobiology: The New Synthesis* in very specific ways and with what would turn out to be key connections to developments in genetics, evolutionary theory and the nature of science debates. Tremendous changes in genetics and biotechnologies were beginning to strengthen the idea that soon, and with enough information about the human genome, we would finally "understand" human nature and be able to offer genetic and attendant biological explanations for not just disease, but complex human behaviors and social problems as well. Work like Wilson's, Dawkins', and Dennett's gained legitimacy because of the increasing biologization of our understanding of human nature and the human condition.

The successes in genetic biotechnologies in the late twentieth century led to the dominance of the idea within genetics that all facets of life including human behaviors can be understood and explained by understanding their genetic and subsequent biological influences. This position included the acceptance, or at least elevation, of a simplistic Mendelianism and an equally limiting adaptationism or what critics would call ultra-Darwinism. This also has been an explicit part of sociobiological discourse and related discourses that have followed. It has also included a process of "trait-fixing," which allows any behavior that can be observed to be considered as arising from selection pressures acting directly on genes. In this process the things that social scientists study, such as complex human behaviors (for example, homosexuality, rape, mating behaviors, and incest) become in large measure epiphenomenal and indicative of processes happening at the genetic and related biological levels.

Likewise, the timing of the proliferation of sociobiological discourse has coincided with the development of a significant and controversial debate about the nature of science and a debate about how we should understand the claims of science. On one side of this debate are those who see science as an entirely objective enterprise capable of producing at least accurate information, if not truth, about the world. On the other side are those who see science as socially constructed and somehow partial and incomplete knowledge influenced by things such as political ideologies and self-interest. The former is the older and more established understanding of science. The latter is the more recent and is highly controversial and ranges along a continuum from moderately conservative to extremely relativist in epistemological orientation.

Wilsonian sociobiological discourse placed itself squarely on the side of those who see science in the more traditional empiricist light. For them, reductionism is not a weakness. It is a strategy of investigation and explanation that is perfectly acceptable and indeed preferred. If human behavior has strong and ultimately determining genetic influences created through adaptation to eons of selection pressures, then we do not need to worry about the things that social scientists study. A particularly extreme version of this position is summed up by Watson, a molecular geneticist and one of the co-discoverers of the helical shape of DNA in an aside to Stephen Rose. Watson remarked: "There is only one science, physics: everything else is social work" (Rose, 1997, 8).

Critics have seen the world very differently. For them, reductionism is not a strategy appropriate to the study of the complex behaviors of man and higher mammals. It tends to destroy the integrity of the levels of reality that social scientists study. It is a practice influenced by what Allen (2002) calls "mechanistic materialism." It renders the social, political, cognitive, and existential levels invisible. Critics and those in the general debate about the nature of science have been wrestling with how to understand science claims. Some want to preserve the sense that science is a unique project, with a strong commitment to materialist and empirical investigation. Others feel that science, as with any other human social project, is so socially overdetermined that its claims are not likely to be more "true" than any other kinds of descriptions of reality, such as speculative philosophy and religion.

A middle ground is also emerging in this debate that sees science as producing socially constructed and partial knowledge. It cannot produce truth, only partial or proximate accuracy. Legitimacy for one's work cannot derive simply from reference to method. Scientists must employ a more modest epistemology that recognizes the diversity of standpoints that can be brought to bear on a research question (Longino, 2002; 2000). Such a modest epistemology recognizes the diverse and equally important levels of reality that must be thoroughly engaged when considering the determinants of human behavior (ibid). The encouraging and teaching of alternative formulations are therefore crucial, since objectivity is not the domain of individual researchers but a societal construction born of pluralism. To this end, Kellert et al. hold that societies must be committed to the ideal of objectivity as a plurality of voices and standpoints (Kellert, Longino, and Waters, 2006). Of equal importance is the recognition that reductive methodologies provide limited knowledge, and alternative formulations need also to include the development of more complex modes of analysis and representation (McKinnon and Silverman, 2005).

In Part II, I have examined selected biology texts to see if and how sociobiology was being presented. I also examined the presentations of genetics and evolutionary theory, and the nature of science, to see if there were correlations between the ways sociobiology was being portrayed and the ways that genetics, evolutionary theory, and the nature of science were also presented. These results are summarized in two sections below, and are tabulated at the end of the chapter in Table 9.1. The four textbooks written in the mid-1990s are presented first and then the two more recent editions of two of the 1990s texts. This allows for the development of a longitudinal perspective and an analysis of shifting patterns in the presentations over time.

The Late Twentieth-Century Textbooks

I have found in this group of textbooks that there is a continuum of presentations of sociobiology from the most unreflectively pro-Wilsonian to the much more carefully nuanced. An important dimension in these presentations was the extent to which the authors did or did not make careful distinctions between human and nonhuman sociobiology. I also found that the presentations of sociobiology in these texts correlated with these texts' presentations of genetics and evolutionary theory. The most pro-Wilsonian presentation of sociobiology is in *Understanding Biology* by Raven and Johnson. This text spends little time presenting evidence for environmental influences on behavior and engages in "mentioning" to appear to cover this terrain. This text also used images as a way to slide the focus from nonhuman to human behaviors, thereby blurring the distinctions and allowing work in nonhuman sociobiology to legitimize the application of the same principles to human behavior. Raven's and Johnson's presentation also correlated with canonical accounts of Mendel and Darwin and simplistic presentations of the gene/behavior dyad, and at the same time they clearly aimed for canonical positions for these scientists and their work within biology. This text did have a section in Chapter 1, called "The Nature of Science", and it did explain to readers that science theory was always subject to change. Overall, however, it presented a traditional view of the nature of science.

Biology: Discovering Life, by Levine and Miller, was in many ways the most careful and nuanced of this group in its claims for human sociobiology. The authors were careful to qualify their discussion of genetics and evolutionary theory, for the most part, and only occasionally endorsed a strong genetic interpretation. For example, in the sections on evolution, they discuss Gould's and Eldredge's theory of punctuated equilibrium that was formulated to counter more strongly adaptationist formulations. It was also the textbook that consciously presented controversial issues as they relate to biology and behavior. It is unclear, however, if this would prove successful with students because, though they discuss many aspects of the debate, they do not provide the reader with a systematic framework for thinking about controversy. In the absence of well-articulated alternative positions, it is unclear if this will leave the reader stimulated and engaged or merely confused. This text also presents a traditional view of the nature of science. It appears at points to move toward a more constructionist position, but in the end upholds the idea that science practice eventually corrects false conclusions. At the same time, this text does make a great effort to present a science and technology-studies perspective that connects scientific theory with technologies and technological problems that develop out of this theory.

Biology: Principles, Patterns, and Processes included a moderate presentation of sociobiology and the relationship between biological and social influences on behavior. This is the only Canadian text and appears to have been "written by committee." This may be why it often seemed very contradictory in its attempt to present balance. So, while it is useful to expose students to both sides of a controversy (much as *Biology: Discovering Life* does) this attempt at a middle ground or neutrality likely serves only to leave students confused. This is especially true if there is no space in the text for teaching about how to consider and analyze controversy in science. Also, this text has no direct discussion of the nature of science, but indirectly in the discussions of Mendel's work it makes allusion to the issue in a contradictory manner. It treated Mendel as a canonical figure, but then at other points it did present his scientific discoveries in context. Students will have difficulties if, at one point, the text seems to endorse, for example, more constructivist positions and then a few paragraphs later argues the opposite position.

Biology (4th edition) by Campbell was the most balanced textbook overall, in this group, in its presentation of sociobiology. It presented the key theoretical basis and also introduced the concept of behavioral ecology. At the same time, it clearly separated the nonhuman and human realms and clearly presented the controversies that swirl around human sociobiology. This was accomplished by presenting the nature/nurture debates as they occurred in the presociobiological field of ethology. This was done well while presenting both the nature and nurture positions thoroughly and clearly. It appeared that Campbell was striving for the contemporary position that biological and environmental influences are each essential and inseparable. Campbell also does not present a simple Mendelianism in the sections on genetics or an ultra-adaptationist position in the sections on evolutionary theory. While this text does have the most sophisticated discussion of the nature of science in this group, it still does not discuss the issue of the social context of scientific discovery. Importantly, however, Campbell presents a sophisticated alternative to reductive formulations in general by introducing the idea of systems biology and the idea of emergent properties within biological systems. This position is further solidified in his presentation of McClintock's work and his encouragement for students to develop "a feel for organisms" rather than just for parts of organisms.

Useful and critical evaluations of any controversial and socially significant science claims, such as those of human sociobiology, need a conceptual and an epistemological foundation. Overall, I have found this foundation lacking in the texts that I examined in this group. None of the texts engaged the issues of epistemology or the nature of science. This is not surprising, given their publication dates. This is important, however, because in the cases of *Biology: Discovering Life* and *Biology: Principles, Patterns, and Processes*, in which the authors attempt to present controversial issues, this lacuna will likely leave the reader confused and lacking a way to make evaluative judgments about the merits of all sides in the controversy.

The Two Twenty-First Century Textbooks

The eighth edition of *Biology*, based on the earlier work of Raven and Johnson, is a much different text than the third edition of *Understanding Biology*. Whereas the third edition Raven and Johnson text was the most supportive of Wilson's sociobiology project, sociobiology is not mentioned in any of the discussions of behavior and genetics in this recent edition. This could mean one of two things. It could mean that the principles promoted by sociobiology have been internalized in the texts as canon and that by removing reference to sociobiology and to Wilson the authors avoid reference to the sociobiology debate and thus sanitize and legitimize the ideas by removing the historical controversy. Or it could mean that the authors have taken what they deem to be useful principles from sociobiology and the ensuing debate and represented these in less controversial and more considered fashion. The latter seems to be the case. While reference is made in the animal behavior sections to Haldane's and Hamilton's work, the authors take pains in these sections and throughout the book to raise the importance of considering biological and environmental factors in development and behavior. They do not succumb to either simple Mendelian or to ultra-adaptationist arguments. Also, only once do they raise the issue of genetic influences on human behavior, and they limit their comments to highlighting the "contested" nature of the topic. On the topic of the nature of science, this edition has a qualified traditional approach, wherein social influences are seen to operate; but in the end, the methods of science and the vigilance of the scientific community are seen to be adequate checks against these social influences. In this edition, as was the case with the Campbell textbooks, reductionism is presented as a method of important but limited value. The authors instead promote a systems biology approach that emphasizes the idea of treating living systems as having emergent properties and synergistic interactions.

Biology (7th edition), written by Campbell and Reece, is, from the perspective of this analysis, the most interesting of all the textbooks examined. For example, it is the only textbook that refers to constructivist theory in science education, and in so doing the authors also refer to debates that have swirled around this pedagogical framework. Also, when considering the influences of nature and nurture, the authors take a combinatorial approach in ways that do not merely "mention" either position. As is the case with the eighth edition of *Biology* discussed above, sociobiology does not have a prominent place in the discussions of behavior and is relegated to a mention in the context of the controversy surrounding conceptions of the determinants of human behavior. In the sections on genetics, the authors avoid simplistic Mendelian formulations, and in the chapters on evolutionary theory, the authors directly undermine an ultra-adaptationist position. Most importantly, the discussions on the nature of science are the most sophisticated of any of the textbooks and include a serious discussion of the advantages and disadvantages of reductionism as a mode of investigation and a systems approach to biological complexities. Both of these newer textbooks have clearly been influenced by sociobiology, the sociobiology debates, recent developments in systems biology,[1] and by recent work done in science education around both constructivist theory and the nature of science. These are important developments and they bode well for both teachers and students.

An important issue for further study that emerged in Part II of this book is the ambiguity in the textbooks around the conceptualization of the terms "character" and "trait." This is clearly not just an issue for textbook authors as it signals a vagueness and circularity in the formulations of these concepts in biology in general. It is important not to see this as just an insignificant academic issue. One of the key steps in reductionist discourse that I discussed on Part I of this book was the ability to "trait fix." This is a multistage process that begins by deeming any behavior a trait and then processing this trait through a simple Mendelian genetics and an ultra-adaptationist evolutionary theory. This process attempts to remove complex social categories and social influences from the behavior and render it an appropriate object of purely biological investigation. Critics have pointed this out and have attempted to critique and deconstruct this process, but this critique is hampered in part by the vagueness of the conceptualizations of the terms "character" and "trait." Part of the critique of "trait fixing" must be to clarify the definitions of these terms in such a way that the social dimensions of behavior are always accounted for and included and the ability to reduce complex behaviors to biological phenomena is restrained.

Implications for Critics

As indicated in the Introduction to this work, the name "sociobiology" has fallen from favor, and this appears to be borne out in the most recent textbooks I examined. If sociobiology and especially human sociobiology, as a "brand," has been removed from biology textbooks, one might ask: Why bother to continue to discuss the issues and arguments it raised? There are many reasons to continue this struggle. The debate over the influences of biology and environment on human behavior has shifted disciplines and is now more commonly found within the terrain of fields such as human behavioral ecology (Allen, 2004, 287–290) and evolutionary psychology (Lewontin and Levins, 2007, 59–63; McKinnon, 2005; Kaplan and Rogers, 2003). It is important to continue to discuss formulations of genetic determinism because some formulations within these newer disciplines share similar conceptions with sociobiological discourse and employ similar arguments and emphases. Equally important, "gene talk" about the determinates of behaviors such as sexual orientation and criminality (Schifellite, 2008) continues to engage the popular imagination, even though more temperate conceptualizations continue to emerge.

At the same time, critics also must continue to develop alternatives to these kinds of arguments. It has not been enough to label these determinisms "pseudoscience," "capitalist science," "patriarchal science," or "race science." These labels have not been enough to undermine the credibility and legitimacy of the ongoing project of promoting biodeterminist arguments about human behavior (Allen, 2004). Practicing scientists, social scientists, and teachers must work cooperatively to develop and encourage debate and to promote alternative conceptions. In this I would echo Longino, in her call for modest epistemological positions and scientific pluralism; Ruse, in his call to respect both good science and cultural influences; and Fausto-Sterling and Harding, in their call for "nonhierarchical, multidisciplinary teams" of researchers who understand the limits of their own disciplines (Fausto-Sterling, 2000a, 255). Yet another reason to continue with this work is that it is also likely that, with continued genetic research and biotechnological developments, we will increasingly be called upon to consider decisions around eugenic interventions (Allen, 2001). This new "liberal" or consumer-based eugenics movement is likely to appear in the form of consumer decisions rather than state mandated practices. Nonetheless, the claims of some of these eugenic biotechnologies likely will be grounded in the same kinds of determinist formulations as has human sociobiological discourse.

Implications for Researchers, Science Educators, and Textbook Authors

These textbook analyses point the way toward what must be done to create both textbooks and teaching practices that help students develop critical skills that can help them both to evaluate controversial issues in science in general and to assess the claims of genetic determinist arguments in particular. Teachers and students of biology, as well as scientists and the general public, need to have an appreciation of the complexities invoked by the sociobiology debate. In short, we need to continue to educate both future scientists and the public about the issues and flaws that surround human genetic determinist arguments. In so doing, it is essential for teachers, students, and the general public to recognize how legitimacy for this project is built. It is also important to present alternative conceptions to human genetic determinist arguments that include serious and nuanced presentations of genetic and evolutionary theory.

It is difficult to fault all of the authors of the earlier texts that I examined for not engaging in clearer counter-hegemonic debates with Wilsonian sociobiology, because alternative conceptions have not yet gained widespread acceptance (Kaplan and Rogers, 2003). This situation is changing, however. It is clear that in addition to the sociobiology debate itself, the completion of the human genome project, advances in molecular biology and developments in systems biology have all had an impact on the two more recent textbooks I analyzed. Through these developments in biology there is an increasing recognition of the systemic complexities at the molecular level that organize life. The authors of these recent textbooks are more sensitive to concerns around reductionism in general and as it applies to formulations that speculate on the determinates of human behavior. The authors of the two more recent textbooks also appear to have been influenced by some of the debates around the nature of science as evidenced by their textbooks. Both these texts provide readers with much more nuanced and in depth arguments around the nature/nurture issue and around the nature of science.

In part, more nuanced and balanced presentations will become more possible as teachers become more familiar with the alternative conceptions that are emerging and as textbooks include presentations of these alternatives. At the same time, we, as a society, must develop ways of fostering debate and a plurality of positions around scientific controversies. In conjunction with teaching students how to evaluate contradictory and controversial positions, both teachers and textbooks must be able to present issues concerning the nature of science and scientific knowledge. To do this, science educators must become familiar with the larger debates on the nature of science and with the larger debates on the nature of organisms. This is a multilevel analysis, because issues around the nature of science emerged as part of the sociobiology debate and also emerged as part of the debate on the nature of science within science studies and science education. Of equal importance, presentation of scientific controversies, presentations of alternative conceptions of the relationship between biology, the environment and behavior, and nuanced presentations of the nature of science can also invoke controversial political issues such as gender and racial and ethnic inequality. Teachers must be well trained to handle and direct these discussions in ways that do not silence participants (Blades, 1997; Hodson, 1994; Van Rooy, 1999; Pedretti, McLaughlin, MacDonald and Gitari,

2001). Textbooks are, and will continue to be, an important site for these debates.

There is a need for more research on the best ways for textbooks to present controversy in science and on the best ways to present discussions about the nature of science itself. This must also include more research to develop pedagogy that fosters students' abilities to understand and conceptualize modest and pluralist epistemological frameworks (Schifellite, 2011; 2002). A study of biology textbooks by Gibbs and Lawson (1992) found that "discussions of the processes of science and scientific thinking occupy a very small segment of most textbooks" (151). They also found that any discussion that does occur is not well integrated with other areas of the text and that generally biology is portrayed as "a collection of facts" (151). This corroborates my findings and it is crucial that teachers begin to demand more coverage of the nature of science in science textbooks and that this coverage be extended throughout the text.[2] An example of an important and interesting textbook study was done by P. Gardner (1999) on science-technology relationships in Canadian physics textbooks. He found that, despite the fact that all six books showed the human influences on the development of physics, all the texts supported the dominant idealist notion that technology follows from the application of science theory. This is against what he calls a more materialist interpretation, in which theory and practice are both necessary and both influence each another. In this research, P. Gardner argues in favor of texts that "reflect modern scholarship about the nature of those relationships" (344). In this he echoes MacDonald's (1996) study of teacher knowledge about evolutionary theory and his calls to make these topics more explicit in biology textbooks and in biology teacher training. Another important area of research to be done on textbooks is to interview textbook authors, with the aim of understanding both the constraints on what and how issues can be presented in textbooks and the explicit aims of authors that are embodied in textbook representations.

Finally, in the context of these discussions, it is important to recognize that much goes on in the interaction of teachers and students that cannot be understood by studying what is presented in textbooks. Although it was beyond the scope of this work, it is crucial for researchers to sit down with teachers and students who are struggling, or have struggled, with these issues and ask them how they make sense of the nature of science and the nature of organisms. Likewise, it would be useful to know how students interact with and interpret information presented in science textbooks. All of this would, I think, be useful for teachers who want to talk about the difficulties they face in presenting this material and the things that would be of use to them in their work. It would also, I think, be important to understand how and in what ways students construct their interpretations of biodeterminist claims they encounter. More of this kind of research can only help enhance the expansion and ultimately the "efficacy" of the debates around the nature of science and organisms.

Notes

1. For an overview of this discipline see, for example, Kitano, 2001; Klipp et al., 2005; Alberghina and Westerhoff, 2005; and Noble, 2006.
2. An interesting early experiment in engaging biology students in discussions about the social nature of science can be seen in *The Study of Biology* (Baker and Allen, 1982). In this text, the authors strove to elucidate the dimension of science as process.

Table 9.1 Comparative Summaries of the Textbooks Analyzed by Theme

Textbooks	Presentations of Sociobiology and Animal Behavior	Presentations of Genetics and Behavior	Presentations of Evolutionary Theory	Presentations of the Nature of Science	Other Presentation Issues
Biology: Discovering Life, 2nd edition by Levine and Miller, 1994	-Nuanced discussion of sociobiology -Well-balanced nature/nurture discussion -Ethology included	-Circular definitions of "character" and "trait" -Mixed messages on genes/environment influences	-Nuanced presentation -Not adaptationist or selectionist -Important discussion on race and genetics	-Science as a social process - More nuanced neopositivist position -Mixed messages on Darwin & Mendel	-Science, technology, and society focus -Use of controversy as pedagogic tool -Likely to be confusing to students
Biology: Principles, Patterns, and Processes by Galbraith, 1989	-Pro sociobiology -Sociobiology debate referenced -Important discussion of the role of learning -Overall mixed messages	-Balanced discussion of interactions of genes/environment -Discussed eugenics and the IQ controversy	-Mixed messages on the importance of selection and adaptation -Covered cytoplasmic inheritance	-No formal discussion -Important social context in discussions of Darwin and Mendel -Nuanced neopositivist position	-Canadian text -Group authorship -Some contradictory positions -Possibly confusing for reader
Understanding Biology, 3rd edition by Raven and Johnson, 1995	-Most supportive of sociobiology -Little discussion of human vs. nonhuman - Strong genetic bias	-Strongly Mendelian -No clear definition of "trait" or "character" -Little on epigenetic influences	-Evolution given hegemonic position -Strong adaptationist and selectionist positions	-Not well developed -Neopositivist -Canonical treatment of Darwin and Mendel -No constructionist issues raised	-"Mentioning" as only real attempt to cover "nurture" issues -Use of photos to "slide" from nonhuman to human examples
Biology, 8th edition based on Raven and Johnson by Losos, Mason, and Singer, 2008	- No mention of sociobiology -Only one example of human behavior -Balanced account of nature/nurture issues	-No simple Mendelianism -Circular definitions of "character" and "trait"	-Evolution given hegemonic position -Much more nuanced & limited adaptationist and selectionist position	-Reference to Systems Biology -More nuanced neopositivist position -Mendel's work is socially qualified	-Different authors from the 3rd edition -Generally much more nuanced presentations than the 3rd edition
Biology, 4th edition by Campbell, 1996	-Ethology & Behavioral ecology both presented -Nuanced nature/ nurture discussion -Nuanced discussion of human sociobiology	-Clearer definitions of "character" and "trait" -Covered cytoplasmic inheritance -No simple Mendelianism	-Population-level focus -Nuanced and limited adaptationist and selectionist positions	-Systems biology presented -Science as process -Darwin discussed in historical context	-Focus on the level of the organism -Interview with E.O. Wilson
Biology, 7th edition by Campbell and Reece, 2005	-Only human sociobiology discussed -Best coverage of the sociobiology debate -Focus on behavioral ecology and learning	-Points out "trait-character" confusion -No simple Mendelianism -Overtly anti-reductionist	-Very nuanced and limited adaptationist and selectionist position -Nonteleological presentation	-Systems biology focus -Social constructionist issues discussed -Balanced discussion of reductionism	-Second author added -Constructivism as pedagogy discussed -Coverage of gender in science issues

APPENDIX I

Former Ontario Ministry of Education Circular 14, 1997, Interim Approved List of Textbooks for Ontario Academic Courses in Biology

1. *Biology*, 3rd ed. Neil A. Campbell. Benjamin/Cummings Publishing, 1993. 1190 pages.
   Interim approval only; see preliminary, section 7(c).
   OAC: Science: Biology
   Hard Cover $61.75 ISBN: 0–8053–1880–1 Circ14# 8239 Available from: Addison-Wesley (416/447–5101)

2. *Biology*, 5th ed. John W. Kimball. Addison-Wesley, 1983. 974 pages.
   Interim approval only; see preliminary, section 7(c).
   OAC: Science: Biology
   Hard Cover $40.10 ISBN: 0–201–10245–5 Circ14# 6003 Available from: Addison-Wesley (416/447–5101)

3. *Biology Directions*. Don Galbraith et al. John Wiley, 1993. 716 pages.
   Interim approval only; see preliminary, section 7(c).
   OAC: Science: Biology
   Senior: Science: Biology, Grade 11
   Hard Cover $54.95 ISBN: 0–471–79512–7 Circ14# 8213 Available from: Nelson (416/752–9100)

4. *Biology: Discovering Life*, 2nd ed. J. S. Levine and K. R. Miller. D. C. Heath Canada, 1994. 898 pages.
   Interim approval only; see preliminary, section 7(c).
   OAC: Science: Biology
   Hard Cover $71.05 ISBN: 0–669–33494–4 Circ14# 8633 Available from: Nelson (416/752–9100)

5. *Biology: Evolution, Diversity, and the Environment*, 2nd ed. Sylvia S. Mader. Wm. C. Brown Publishers, 1987. 772 pages.
   Interim approval only; see preliminary, section 7(c).
   OAC: Science: Biology
   Hard Cover $36.00 ISBN: 0–697–01357-X Circ14# 6560 Available from: Asquith House (416/925–3577)

6. *Biology: Exploring Life*, 2nd ed. G. Brum, L. McKane, and G. Karp. John Wiley, 1994. 1026 pages.
   Interim approval only; see preliminary, section 7(c).
   OAC: Science: Biology
   Hard Cover $48.95 ISBN: 0–471–54408–6 Circ14# 8720 Available from: John Wiley & Sons Canada (416/236–4433)

7. *Biology: Life on Earth*, 3rd ed. G. Audesirk and T. Audesirk. Macmillan Publishing, 1993. 1049 pages.
   Interim approval only; see preliminary, section 7(c).
   OAC: Science: Biology
   Hard Cover $56.10 ISBN: 0–02–304811–5 Circ14# 8214 Available from: Maxwell Macmillan (416/449–6030)

8. *Biology: Living Systems*, 7th ed. Raymond F. Oram. Glencoe/Macmillan/McGraw-Hill, 1994. 966 pages.
   Teacher's resource book available.
   Interim approval only; see preliminary, section 7(c).
   OAC: Science: Biology
   Senior: Science: Biology, Grade 11
   Hard Cover $57.89 ISBN: 0–02–800672–0 Circ14# 8536 Available from: McGraw-Hill Ryerson (905/430–5050)

9. *Biology: Principles, Patterns, and Processes*. Don Galbraith et al. John Wiley, 1989. 656 pages.
   OAC: Science: Biology
   Hard Cover $43.95 ISBN: 0–471–79629–8 Circ14# 7155 Available from: Nelson (416/752–9100)

10. *Biology: A Canadian Laboratory Manual*. R. Ritter, B. Drysdale, and D. Coombs. GLC/Silver Burdett Publishers, 1987. 250 pages.
    OAC: Science: Biology

Soft Cover $17.50 ISBN: 0–88874–039–5 Circ14# 6266 Available from: GLC/Silver Burdett Publishers (416/497–4600)

11. *Biology: The Study of Life*, 3rd ed. W. D. Schraer and H. J. Stoltze. Allyn & Bacon, 1990. 766 pages.
    Interim approval only; see preliminary, section 7(c).
    OAC: Science: Biology
    Senior: Science: Biology, Grade 11
    Hard Cover $62.02 ISBN: 0–13–080681–1 Circ14# 7156 Available from: Prentice-Hall Canada (416/293–3621)

12. *Biology: The Unity and Diversity of Life*, 7th ed. C. Starr and R. Taggart. Wadsworth Publishing, 1995. 933 pages.
    Teacher's resource book available.
    Interim approval only; see preliminary, section 7(c).
    OAC: Science: Biology
    Hard Cover $66.56 ISBN: 0–534–21060–0 Circ14# 8537 Available from: Nelson (416/752–9100)

13. *Nelson Biology*, rev. ed. Bob Ritter et al. Nelson, 1993. 716 pages.
    Teacher's resource book available.
    OAC: Science: Biology
    Senior: Science: Biology, Grade 11
    Hard Cover $54.95 ISBN: 0–17–603870–4 Circ14# 8281 Available from: Nelson (416/752–9100)

14. *Understanding Biology*, 3rd ed. P. H. Raven and G. B. Johnson. Wm. C. Brown Publishers, 1995. 872 pages.
    Teacher's resource book available.
    Interim approval only; see preliminary, section 7(c).
    OAC: Science: Biology
    Hard Cover $45.75 ISBN: 0–697–22213–6 Circ14# 8603 Available from: Asquith House (416/925–3577)

APPENDIX II

Notes About Method: Textual Analysis as Qualitative Research

Situating the Researcher

The two kinds of textual analyses in this book are both forms of qualitative research; in both cases, validity rests on my accurately representing the ideas presented in the texts and my making the process of interpreting those ideas as transparent as possible. I do not try to claim neutrality in the sense of not introducing bias into the process of interpretation. It is understood that this is an active process of interpretation. Rather, it is important that the researcher's biases be as clearly articulated as possible, so as to make the process of interpretation transparent and understandable. Clearly, my work on the sociobiology debates, my position on human sociobiology, and my previous work in ideology studies all have influenced my choice of texts and my focus in this textual analysis. To this end, I have tried to make my own opinions on the sociology debate and on the nature of science controversy visible in the preceding chapters. I also have clearly indicated detailed references where I have cited the texts in order to make my interpretations traceable.

Collecting Data and Analyzing Themes

Issues in the Development and Presentation of this Work

The presentation of research findings appears to have an ordering that is usually quite different from the actual processes and ordering of events of the research itself. It is difficult to establish a linear progression for the development of my ideas in this work. This textual analysis project has grown in ways that I understand to be consistent with the epistemological foundations of qualitative work. I began with one question: "Is sociobiology being covered in specific introductory biology texts and if so, in what manner?" A second question then arose: "How are themes that arose in the human sociobiology debates being covered in the textbooks?" And then a third question arose: "How do these texts cover scientific controversies in general?"

I had done a good deal of reading on the first twenty-five years of the sociobiology controversy prior to doing the analysis of the textbooks. I had begun to form opinions about what I thought of the whole project. I was critical and I felt that how one thought about human sociobiology was strongly correlated with the kind of genetic theory, evolutionary theory, and general epistemological frameworks that one also held. I knew that I wanted to look at not just what the textbooks said about human and nonhuman sociobiology but also at what they said about behavior in general and how they presented genetics, evolutionary theory, and what has come to be called "the nature of science."

As a qualitative researcher, I have been concerned with articulating what I feel are the general tendencies, nuances, interrelations, and subtexts implicit in the textbook presentations. I have gone about collecting data by first examining the long and short forms of the tables of contents in each text and taking note of the four areas or themes identified above. With a good sense of the overview and subheadings in each text, I then began a detailed process of reading and taking notes for each text.[1] After all notes were taken, I then went back over my notes and pulled out themes that seemed important to my work.[2]

In the re-examination of the sociobiology debate, the re-examination of my notes taken from the texts, and the articulation of themes, I developed ideas that later appear to precede the investigation but actually do not. My research on the sociobiology debate doubtless influences what I see in the texts, and what I see in the texts doubtless influences what I see in the sociobiology debates. I did not, in fact, do one before the other but rather first worked on the debates and then examined the texts and then re-examined the debates and re-examined the work I had done on the texts. Upon completing Part I of this book, my focus sharpened. As I began to analyze the texts and to further analyze the sociobiology debate, I came to see that I could not separate the presentations of sociobiology from presentations of other related disciplines, especially genetics and evolutionary theory. I also came to see that in order to do this kind of analysis, I needed a position on the nature of science debates as well. These issues, in effect, emerged as the themes in my textual qualitative research.

Notes

1. I found this exercise to be very enjoyable. All these textbooks contained many fascinating and clearly developed presentations. It rekindled in me that awe and fascination with the wide range of topics covered in a general biology text. This was, for me, reminiscent of my Introductory Biology course in university; the marvelous text *Biological Science*, by Keeton (1967), spurred me on to a two-major Bachelor of Science degree in biology and psychology.
2. Qualitative researchers generally go through numerous iterations and approximations of the data they collect in the process of organizing this data into sensible and hierarchical conceptual levels. For a more extensive discussion, see Neuman, 2000.

BIBLIOGRAPHY

Abd-El-Khalick, F., R. L. Bell, and N. Lederman. (1998). "The Nature of Science and Instructional Practice: Making the Unnatural Natural," *Science Education*, 82, 417–436.

Abir-Am, P. (1982). "The Discourse of Physical Power and Biological Knowledge in the 1930s: Reappraisal of the Rockefeller Foundation's "Policy" in Molecular Biology". *Social Studies of Science* 12: 341-382.

Aikenhead, G. (1989). "Categories of STS Instruction". *STS Research Network Missive 3:2, 20-23.*

———. (1994a). "What Is STS Science Teaching?" in Solomon, J. and G. Aikenhead, eds. *STS Education: International Perspectives on Reform*. New York: Teachers College Press, 47–59.

———. (1994b). "The Social Contract of Science: Implications for Teaching Science" in Solomon, J. and G. Aikenhead, eds. *STS Education: International Perspectives on Reform*. New York: Teachers College Press, 11–20.

———. (1998). *Teaching Science Through a Science-Technology-Society-Environment Approach: An Instruction Guide* (SIDRU Research Report 12). Saskatoon, Saskatchewan: University of Regina, Saskatchewan Instructional Development and Research Unit.

Alberghina, L. and H. Westerhoff, eds. (2005). *Foundations of Systems Biology: Definitions and Perspectives*. Berlin Heidelberg: Springer-Verlag.

Alcock, J. (1975). *Animal Behavior: An Evolutionary Approach*. Sunderland, MA: Sinauer Associates.

———. (2001). *The Triumph of Sociobiology*. Oxford and New York: Oxford University Press.

Alexander, R. D. (1977). "Natural Selection and the Analysis of Human Sociality" in Goulden, C. E., ed. *Changing Scenes in Natural Sciences*. Philadelphia: Philadelphia Academy of Natural Sciences, 283–337.

———. (1979). *Darwinism and Human Affairs*. Seattle: University of Washington Press.
———. (1987). *The Biology of Moral Systems*. Hawthorne, NY: A. de Gruyter.
———. (1988). *Darwinism and Human Affairs*. Seattle: University of Washington Press.
———. (1990). *How Did Humans Evolve? Reflections on the Uniquely Unique Species*. Ann Arbor: Museum of Zoology, University of Michigan.

Allen, E., et al. (1978). "Against 'Sociobiology'" in A. L. Caplan, ed. *The Sociobiology Debate*. New York, Hagerstown, San Francisco, and London: Harper and Row, 259–264. Originally published as "Against 'Sociobiology,'" *The New York Review of Books* (November 13, 1975), 182, 184–186.

Allen, G. (1978). *Life Science in the Twentieth Century*. London and New York: Cambridge University Press.
———. "Ernst Mayr and Biological Thought." *Science and Nature*, 6 (1983), 21–29.
———. (1991). "Old Wine in New Bottles: From Eugenics to Population Control in the Work of Raymond Pearl" in Benson, K. R., J. Maienschein, and R. Rainger, eds. *The Expansion of American Biology*. New Brunswick: Rutgers University Press, 231–261.
———. (1994). "The Genetic Fix: The Social Origins of Genetic Determinism" in Tobach, E. and B. Rosoff, eds. *Challenging Racism and Sexism: Alternatives to Genetic Explanations*. New York: The Feminist Press of the City University of New York, 163–187.
———. (2001). "Essays on Science and Society: Is a New Eugenics Afoot?" *Science* 294 (October 5), 59–61.
———. (2002). "The Classical Gene: Its Nature and Its Legacy" in Parker, L. and R. Ankeny (eds), *Mutating Concepts, Evolving Disciplines: Genetics, Medicine and Society*. Dordrecht, Netherlands: Kluwer Academic Publishers, 11-41.
———. (2004). "DNA and Human-Behavior Genetics: Implications for the Criminal Justice System" in Lazer, David. ed. *DNA and the Criminal Justice System*. Cambridge, MA: MIT Press, 287–314.
———. (2007). "A Century of Evo-Devo: The Dialectics of Analysis and Synthesis in Twentieth-Century Life Science" in Laublichler, M. D. and J. Maienschein, eds. *From Embryology to Evo-Devo*. Cambridge, MA: MIT Press, 123–168.

Altbach, P. G. (1991). "Textbooks: The International Dimension" in Apple, M. W. and L. K. Christian-Smith, eds. *The Politics of the Textbook*. New York and London: Routledge, 242–258.

Apple, M. W. (1991). "The Culture and Commerce of the Textbook" in Apple, M. W. and L. K. Christian-Smith, eds. *The Politics of the Textbook*. New York and London: Routledge, 22–40.

Apple, M. W. and L. K. Christian-Smith (1991). "The Politics of the Textbook" in Apple, M. W. and L. K. Christian-Smith, eds. *The Politics of the Textbook*. New York and London: Routledge, 1–21.

Apple, M. W. and L. K. Christian-Smith, eds. (1991). *The Politics of the Textbook*. New York and London: Routledge.

Aronowitz, S. (1988). Science as Power: Discourse and Ideology in Modern Society. Minneapolis: University of Minnesota Press.

Bachelard, G. (1984). *The New Scientific Spirit*. Boston: Beacon Press.

Badley, J. H. (1931). *The Will to Live: An Outline of Evolutionary Psychology*. London: Allen.

Baker, J. J. W. and G. E. Allen (1982). *The Study of Biology*. 4th ed. Reading, MA: Addison-Wesley.

Baker, R. (2000). *Sex in the Future: The Reproductive Revolution and How It Will Change Us*. New York: Arcade Publishing.

Baldwin, J. D. and J. I. Baldwin (1981). *Beyond Sociobiology*. New York: Elsevier.

Barash, D. P. (1977). *Sociobiology and Behavior*. 2nd ed. New York: Elsevier-North, Holland.

———. (1979). *The Whisperings Within*. New York: Harper and Row.

Barash, D. P. and I. Barash (2000). *The Mammal in the Mirror: Understanding Our Place in the Natural World*. New York: W. H. Freeman and Company.

Barkow, J. H., L. Cosmides, and J. Tooby, eds. (1992). *The Adapted Mind: Evolutionary Psychology and the Generation of Culture*. New York: Oxford University Press.

Barlow, G. W. (1991). "Nature-Nurture and the Debate Surrounding Ethology and Sociobiology." *American Zoologist* 31, 286–296.

Barlow, G. W. and J. Silverberg, (1980). *"Sociobiology: Beyond Nature/Nurture?"* Paper presented at the AAAS Selected Symposium 35, Boulder, CO.

Barnes, B. (1974). *Scientific Knowledge and Sociological Theory*. London: Routledge and Kegan Paul.

Baruch, S., D. Kaufman, and K. Hudson. (2006). "Genetic Testing of Embryos: Practices and Perspectives of U.S. IVF Clinics." *Fertility and Sterility*, 89:5, 1053–8

Bedaux, J. B. and B. Cooke, eds. (1999). *Sociobiology and the Arts*. Amsterdam and Atlanta: Editions Rodopi.

Benson, K. R., J. Maienschein, and R. Rainger, eds. (1991). *The Expansion of American Biology*. New Brunswick, NJ: Rutgers University Press.

Bernal, J. D. (1939). *The Social Function of Science*. London: Routledge.

———. (1971). *Science in History*. Cambridge, MA: MIT Press.

Betta, M. (2006)."From Destiny to Freedom? On Human Nature and Liberal Eugenics in the Age of Genetic Manipulation"in M. Betta ed.*The Moral, Social, and Commercial Imperatives of Genetic Screening: The Australian Case* Dordrecht. The Netherlands: Springer, (3-24).

Bickhard, M. H. (1997). "Constructivism and Relativisms: A Shopper's Guide." *Science and Education* 6 (1–2), 29–42.

Birke, L. I. A. and R. Hubbard, eds. (1995). *Reinventing Biology: Respect for Life and the Creation of Knowledge*. Bloomington: Indiana University Press.

Birke, L. I. A. and J. Silvertown, eds. (1984). *More than the Parts: Biology and Politics*. London and Sydney: Pluto Press.

Blades, D. (1997). "Towards a Post-Modern Science Education Curriculum-Discourse: Repetition of a Dream Catcher." *Journal of Educational Thought*, 31(1), 31–44.

Bloor, D. (1976). *Knowledge and Social Imagery*. London: Routledge and Kegan Paul.

Blute, M. (2003). "The Evolutionary Ecology of Science".*Journal of Memetics - Evolutionary Models of Information Transmission*, 7:1. http://cfpm.org/jom-emit/2003/vol7/blute_m.html

———. (2010). *Darwinian Sociocultural Evolution: Solutions to Dilemmas in Cultural and Social Theory*. Cambridge: Cambridge University Press.

Bocock, R. (1986). *Hegemony*. Chichester [West Sussex], London and New York: E. Horwood and Tavistock Publications.

Boyd, R. and P. J. Richerson (1985). *Culture and the Evolutionary Process*. Chicago: University of Chicago Press.

———. (1990). "Group Selection Among Alternative Stable Strategies".*Journal of Theoretical Biology*, 145, 331 - 342.

Brannigan, A. (1981). *The Social Basis of Scientific Discoveries*. Cambridge [England] and New York: Cambridge University Press.

Brown, A. (1999). *The Darwin Wars: How Stupid Genes Became Selfish Gods*. London: Simon and Schuster.

Brown, J. L. (1975). *The Evolution of Behavior*. 1st ed. New York: Norton.

BSSRS Sociobiology Group (1984a). "Human Sociobiology" in Birke, L. I. A. and J. Silvertown, eds. *More Than the Parts: Biology and Politics*. London and Sydney: Pluto Press, 110–135.

———. (1984b). "Animal Behavior to Human Nature: Ethological Concepts of Dominance" in Birke, L. I. A. and J. Silvertown, eds. *More Than the Parts: Biology and Politics*. London and Sydney: Pluto Press, 136–151.

Burian, R. (1978). "A Methodological Critique of Sociobiology" in Caplan, A. L., ed. *The Sociobiology Debate: Readings on Ethical and Scientific Issues*. New York, Hagerstown, San Francisco, and London: Harper and Row, 376–395.

Burnham, T. and J. Phelan (2001). *Mean Genes: From Sex to Money to Food: Taming Our Primal Instincts*. Cambridge, MA: Perseus.

Burtt, E. A. (1932). *The Metaphysical Foundations of Modern Science*. London: Routledge and Kegan Paul.

Buss, D. M. (1994). *The Evolution of Desire: Strategies of Human Mating*. New York: Basic Books.

———. (1998). *Evolutionary Psychology: The New Science of the Mind*. Boston: Allyn and Bacon.

Campbell, N. A. (1993). *Biology*. 3rd ed. Menlo Park, CA: Benjamin Cummings Publishing Company.

———. (1996). *Biology*. 4th ed. Menlo Park, CA: Benjamin Cummings Publishing Company.

Campbell, N. A. and J. Reece. (2005). *Biology*. 7th ed. San Francisco: Pearson Educational as Benjamin Cummings.

Caplan, A. L. (1978). *The Sociobiology Debate: Readings on Ethical and Scientific Issues*. New York, Hagerstown, San Francisco, and London: Harper and Row.

Capra, F. (1983). *The Turning Point: Science, Society and the Rising Culture*. New York: Bantam.

Caspari, E. (1958). "The Genetic Basis of Behavior" in Roe, A. and G. G. Simpson, eds. *Behavior and Evolution*. New Haven: Yale University Press, 130–127.

Cavalli-Sforza, L. L. and M. W. Feldman (1981). *Cultural Transmission and Evolution: A Quantitative Approach*. Princeton, NJ: Princeton University Press.

Cawthron, E. R. and J. A. Rowell (1978). "Epistemology and Science Education." *Studies in Science Education* 5, 31–59.

Chang, P. (1994). *Marxism and Human Sociobiology: The Perspective of Economic Reforms in China*. Albany: State University of New York Press.

Chase, I. D. and T. DeWitt. (1988). "Vacancy Chains: A Process of Mobility to New Resources in Humans and Animals." *Social Science Information*. 27, 81–96.

Chase, Ivan D., M. Weissberg, and T.H. DeWitt. (1988). "The Vacancy Chain Process: A New Mechanism of Resource Distribution in Animals with Application to Hermit Crabs." *Animal Behaviour*. 36, 1265–1274.

Clark, W. R. and M. Grunstein (2000). *Are We Hardwired? The Role of Genes in Human Behavior*. Oxford and New York: Oxford University Press.

Clough, M. P. (1998). "Integrating the Nature of Science with Student Teaching: Rationales and Strategies" in McComas, W. F., ed. *The Nature of Science in Science Education*. Dordrecht, Boston, and London: Kluwer Academic Publishers, 197–209.

Clutton-Brock, T. H., M. J. O'Riain, P. N. M. Brotherton, D. Gaynor, R. Kansky, A. S. Griffin, et al. (1999). "Selfish Sentinels in Cooperative Mammals." *Science* 284 (5420), 1640–1644.

Cobern, W. W. (1998a). "Science and a Social Constructivist View of Science Education" in W. W. Cobern, W. W., ed. *Socio-Cultural Perspectives on Science Education: An International Dialogue*. Norwell, MA: Kluwer Academic Publishers, 7–23.

Cobern, W. W., ed. (1998b). *Socio-Cultural Perspectives on Science Education: An International Dialogue*. Norwell, MA: Kluwer Academic Publishers.

Cobern, W. W., and C. C. Loving (1998). "The Card Exchange: Introducing the Philosophy of Science" in W. F. McComas, ed. *The Nature of Science in Science Education*. Dordrecht, Boston, and London: Kluwer Academic Publishers, 73–82.

Cohen, J (1989). Propaganda from the Middle of the Road: The Centrist Ideology of the News Media". *Fairness and Accuracy in Reporting*. October/November. http://www.fair.org/index.php?page=1492.

Coll, C. G., E. L. Bearer, and R. M. Lerner (2004). "Nature and Nurture in Human Behavior and Development: A View of the Issues" in Coll, C.G., E. L. Bearer, and R. Lerner, eds. *Nature and Nurture: The Complex Interplay of Genetic and Environmental Influences on Human Behavior and Development*. Mahwah, New Jersey: Lawrence Erlbaum Associates, Inc.

Creath, R. and J. Maienschein, eds. (2000). *Biology and Epistemology*. Cambridge, UK, and New York: Cambridge University Press.

Dahl, E. (2003). "Ethical Issues in the Uses of Preimplantation Genetic Diagnosis." *Human Reproduction* Vol. 18, No. 7, 1368–1369.

Darden, L. (1992). "Character: Historical Perspectives" in Keller, E. F. and E. A. Lloyd, eds. *Keywords in Evolutionary Biology*. Cambridge, MA and London: Harvard University Press, 41–44.

Davis, P. and D. H. Kenyon (1996). *Of Pandas and People: The Central Question of Biological Origins*. Haughton Publishing Company.

Dawkins, R. (1976). *The Selfish Gene*. New York: Oxford University Press.

———. (1979). "Defining Sociobiology." *Nature* 280, 427–428.

———. (1982). *The Extended Phenotype: The Gene as the Unit of Selection*. Oxford [Oxfordshire] and San Francisco: Freeman.

———. (1986). *The Blind Watchmaker*. Harlow: Longman Scientific and Technical.

———. (1989). *The Selfish Gene*. 2nd ed. Oxford: Oxford University Press.

———. (1995). *River out of Eden: A Darwinian View of Life*. London: Weidenfeld and Nicolson.

———. (2006). *The God Delusion*. London: Bantam Press, Transworld Publishers.

———. (2006a). *The Selfish Gene*. 30th Anniversary Ed. New York: Oxford University Press.

Degler, C. N. (1991). *In Search of Human Nature: The Decline and Revival of Darwinism in American Social Thought*. New York: Oxford University Press.

Dennett, D. C. (1995). *Darwin's Dangerous Idea: Evolution and the Meanings of Life*. New York: Simon and Schuster.

Desan, W. (1961). *The Planetary Man*. Vol. 1. Washington, D.C.: Georgetown University Press.

Desan, W. (1972). *The Planetary Man*. Vol. 2. New York: Macmillan.

DeWitt, T. J. and S. M. Scheiner (2004). "Phenotypic Variation from Single Genotypes: A Primer" in DeWitt, T. J., and S. M. Scheiner, eds. *Phenotypic Plasticity: Functional and Conceptual Approaches*. New York: Oxford University Press.

DiLascia, P. (2006). "How Many Children?" *Detroit Free Press*. May 23, 2006.

Dobzhansky, T. G. (1956). *The Biological Basis of Human Freedom*. New York: Columbia University Press.

———. (1964). *Heredity and the Nature of Man*. New York: Harcourt Brace and World.

———. (1967). *Mankind Evolving: The Evolution of the Human Species*. New Haven: Yale University Press.

———. (1970). "Of Flies and Men" in Robinson, D. N., ed. *Heredity and Achievement: A Book of Readings*. New York: Oxford University Press, 425–441.

Dreyfus, H. L. and P. Rabinow (1982). *Michel Foucault: Beyond Structuralism and Hermeneutics*. Brighton, UK: The Harvester Press.

Driscoll, M. P., M. Moallem, W. Dick, and E. Kirby. (1994). "How Does the Textbook Contribute to Learning in a Middle School Science Class?" *Contemporary Educational Psychology* 19(1), 79–100.

Durham, W. H. (1991). *Coevolution: Genes, Culture, and Human Diversity*. Stanford, CA: Stanford University Press.

Durkheim, E. (1915). *The Elementary Forms of the Religious Life*. London: G. Allen and Unwin.

Duschl, R. A. (1985). "Science Education and Philosophy of Science: Twenty-Five Years of Mutually Exclusive Development." *School Science and Mathematics* 85(7) (1985), 541–553.

———. (1990). *Restructuring Science Education: The Importance of Theories and Their Development*. New York: Teachers College Press.

Eagleton, T. (1991). *Ideology: An Introduction*. London: Verso.

Eibl-Eibesfeldt, I. (1975). *Ethology, the Biology of Human Behavior*. New York: Holt, Reinhart and Winston.

Epstein, S. (1996). *Impure Science: AIDS, Activism, and the Politics of Knowledge*. Berkeley, CA: University of California Press.

Etkin, W. (1981). "A Biological Critique of Sociobiological Theory" in White, E., ed. *Sociobiology and Human Politics*. Lexington, MA, and Toronto: Lexington Books, 45–97.

Fausto-Sterling, A. (2000a). *Sexing the Body: Gender Politics and the Construction of Sexuality*. New York: Basic Books.

———. (2000b). "The End of Gene Control." Internet site: www.edge.org.

Femia, J. V. (1987). *Gramsci's Political Thought: Hegemony, Consciousness, and the Revolutionary Process*. Oxford: Clarendon Press.

Feyerabend, P. (1987). *Farewell to Reason*. London: Verso.

Fisher, R. A. (1936). "Has Mendel's Work Been Rediscovered?" *Annals of Science*, 1, 115–137.

Fleck, L. (1979). *Genesis and Development of a Scientific Fact* (Bradley, F., and T. J. Trenn, Trans.). Chicago and London: The University of Chicago Press.

Foucault, M. (1970). *The Order of Things* (Sheridan, A., Trans.). New York: Random House.

———. (1972). *The Archaeology of Knowledge* (Sheridan, A., Trans.). London: Tavistock Publications.

Fristrup, K. (1992). "Character: Current Usages" in Keller, E. F., and E. A. Lloyd, eds. *Keywords in Evolutionary Biology*. Cambridge, MA, and London: Harvard University Press, 45–51.

Fuller, J. L., and W. R. Thompson (1960). *Behavior Genetics*. New York: Wiley.

Fuller, S. (2006). *The New Sociological Imagination*. London: Sage Publications.

Galbraith, D. I. (1989). *Biology: Principles, Patterns, and Processes*. Toronto: J. Wiley.

———. (1993). *Biology Directions*. Toronto: John Wiley and Sons.

Gardner, P. L. (1999). "The Representation of Science-Technology Relationships in Canadian Physics Textbooks." *International Journal of Science Education* 21:3, 329–347.

Gardner, W. (1995). "Can Human Genetic Enhancement Be Prohibited?" *Journal of Medicine and Philosophy* 20, 65–84.

Garry, A., and M. Pearsall (1989). *Women, Knowledge, and Reality: Explorations in Feminist Philosophy*. Boston: Unwin Hyman.

———. (1996). *Women, Knowledge, and Reality: Explorations in Feminist Philosophy*. 2nd ed. New York: Routledge.

Geelan, D. R. (1997). "Epistemological Anarchy and the Many Forms of Constructivism." *Science and Education* 6, 15–28.

Ghiselin, M. T. (1974). *The Economy of Nature and the Evolution of Sex*. Berkeley: University of California Press.

Gibbs, A., and A. E. Lawson. (1992). "The Nature of Scientific Thinking as Reflected by the Work of Biologists and by Biology Textbooks." *The American Biology Teacher* 54(3), 137–152.

Gieryn, T. F. (1999). *Cultural Boundaries of Science: Credibility on the Line*. Chicago: University of Chicago Press.

Goetz, T. (2007). Decode Your DNA. *Wired* 15:12, 256-265.

Gottfried, S. S., and W. C. Kyle, Jr. (1992). "Textbook Use and the Biology Education Desired State." *Journal of Research in Science Teaching* 29(2), 35–49.

Gould, S. J. (1977). *Ever Since Darwin*. New York: W. W. Norton.

———. (1978). "Biological Potential vs. Biological Determinism" in Caplan, A. L., ed. *The Sociobiology Controversy*. New York: Harper and Row, 343–351. Originally published as "Biological Potential vs. Biological Determinism." *Natural History* (May 1976).

———. (1980a). "Sociobiology and the Theory of Natural Selection" in Barlow, G. W., and J. Silverberg, eds. *Sociobiology: Beyond Nature/Nurture?* Boulder, CO: Westview Press for the American Association for the Advancement of Science, 257–269.

———. (1980b). "Sociobiology and Human Nature: A Postpanglossian Vision" in Montagu, A., ed. *Sociobiology Examined*. New York and Oxford: Oxford University Press, 283–290. Originally published as "Sociobiology and Human Nature: A Postpanglossian Vision." *Human Nature* 1:10 (1978).

———. (1981). *The Mismeasure of Man*. New York: W. W. Norton.

———. (1984). "Morphological Channeling by Structural Constraint: Convergence in Styles of Dwarfing and Gigantism in Cerion, with a Description of Two New Fossil Species and a Report on the Discovery of the Largest Cerion." *Paleobiology* 10, 172–194.

———. (1989). *Wonderful Life: The Burgess Shale and the Nature of History*. New York: W. W. Norton.

———. (1993). *Eight Little Piggies: Reflections in Natural History*. New York: W. W. Norton.

Gould, S. J., and R. C. Lewontin. (1979). "Spandrels of San Marco and the Panglossian Paradigm: A Critique of the Adaptationist Programme." Paper presented at the Proceedings of the Royal Society, London.

Gramsci, A. (1971). *Selections from the Prison Notebooks of Antonio Gramsci*. (Hoare, Q., and G. Nowell-Smith, Trans.). London: Lawrence and Wishart.

Grandy, R. E. (1997). "Constructivism and Objectivity: Disentangling Metaphysics from Pedagogy." *Science and Education* 6(1–2), 43–53.

Gregory, B. (1988). *Inventing Reality: Physics as Language*. New York, Chichester, Brisbane, Toronto, and Singapore: John Wiley and Sons, Inc., Wiley Scientific Editions.

Gross, P. and N. Levitt. (1994). *Higher Superstition: The Academic Left and its Quarrels with Science*. Baltimore: Johns Hopkins University Press.

Gross, P., N. Levitt, and M.W. Lewis. eds. (1996). *The Flight from Science and Reason*. New York: New York Academy of Sciences.

Gutting, G. (1989). *Michel Foucault's Archaeology of Scientific Reason*. Cambridge, UK and New York: Cambridge University Press.

Hacking, I. (1982). "Language, Truth and Reason" in M. Hollis and S. Lukes (eds). *Rationality and Relativism*. Oxford: Blackwell, 48-66.

———. (1992). "'Style' for Historians and Philosophers". *Studies in History and Philosophy of Science*, 23:1, 1-20.

Haddad, R. and C. Schifellite, C. (1993). "Patriarchal Conceptions in Use in Modern Virology" in T. Haddad, ed. *Reconstructing Canadian Men and Masculinity*. Toronto: Canadian Scholars Press, 1993, 183–199.

Haila, Y., and R. Levins (1992). *Humanity and Nature*. London: Pluto Press.

Hall, C. (1951). "The Genetics of Behavior" in Stevens, S. S., ed. *Handbook of Experimental Psychology*. New York: Wiley, 304–329.

Hamer, D. H., and P. Copeland (1994). *The Science of Desire: The Search for the Gay Gene and the Biology of Behavior*. New York: Simon and Schuster.

———. (1998). *Living with Our Genes: Why They Matter More Than You Think*. New York: Doubleday.

Hamilton, W. D. (1964). "The Genetical Evolution of Social Behavior: I and II." *Journal of Theoretical Biology* 7, 1–52.

———. (1996). *Narrow Roads of Gene Land: The Collected Papers of W. D. Hamilton*. Oxford, UK, and New York: W. H. Freeman.

Hammerich, P. L. (1998). "Confronting Students' Conceptions of the Nature of Science with Cooperative Controversy" in McComas, W. F., ed. *The Nature of Science in Science Education*. Dordrecht, Boston, and London: Kluwer Academic Publishers, 127–136.

Hanson, N. R. (1958). *Patterns of Discovery*. Cambridge: Cambridge University Press.

Haraway, D. J. (1976). *Crystals, Fabrics, and Fields: Metaphors of Organicism in Twentieth-Century Developmental Biology*. New Haven, CT: Yale University Press.

———. (1989). *Primate Visions: Gender, Race, and Nature in the World of Modern Science*. New York: Routledge.

———. (1991). *Simians, Cyborgs, and Women: The Reinvention of Nature*. London: Free Association Books.

Harding, S. G. (1986). *The Science Question in Feminism*. Ithaca, NY: Cornell University Press.

———. (1991). *Whose Science? Whose Knowledge?: Thinking from Women's Lives*. Ithaca, NY: Cornell University Press.

———. (1998). *Is Science Multicultural? Postcolonialisms, Feminisms, and Epistemologies*.

Bloomington, IN: Indiana University Press.

Harding, S. G., and M. B. Hintikka. (1983). *Discovering Reality: Feminist Perspectives on Epistemology, Metaphysics, Methodology, and Philosophy of Science*. Dordrecht, Holland and Boston: D. Reidel.

Hausfater, G., and S. B Hrdy (1984). *Infanticide: Comparative and Evolutionary Perspectives*. New York: Aldine Publishing Company.

Herman, E. S., and N. Chomsky (1988). *Manufacturing Consent: The Political Economy of the Mass Media*. New York: Pantheon Books.

Herrick, C. J. (1963). *Brains of Rats and Men: A Survey of the Origin and Biological Significance of the Cerebral Cortex*. New York: Hafner Publishing Company.

Herrnstein, R. J., and C. A. Murray (1994). *The Bell Curve: Intelligence and Class Structure in American Life*. New York: Free Press.

Herron, J. P., and A. G. Kirk, eds. (1999). *Human/Nature: Biology, Culture, and Environmental History*. Albuquerque: University of New Mexico Press.

Herzog, Jr., H. A. (1986). "The Treatment of Sociobiology in Introductory Psychology Textbooks." *Teaching of Psychology*13:1 (February), 12–15.

Hessen, B. (1931). "The Social and Economic Roots of Newton's 'Principia.'" In *Science at the Crossroads: Papers Presented to the International Congress of the History of Science and Technology, Held in London from June 29th to July 3rd, 1931 / by the Delegates of the U.S.S.R.* (147–212). Aldwych, London: Kniga (UK) Ltd. Bush House.

Hinde, R. A. (1974). *Biological Bases of Human Social Behaviour*. New York: McGraw-Hill.

Ho, M. W. (1998). *The Rainbow and the Worm: The Physics of Organisms*. 2nd ed. Singapore and River Edge, NJ: World Scientific.

Hodson, D. (1988). "Toward a Philosophically More Valid Science Curriculum." *Science Education* 72:2, 19–40.

Hodson, D. (1991). "Philosophy of Science and Science Education" in Matthews, M. R., ed. *History, Philosophy, and Science Teaching: Selected Readings*. Toronto: OISE Press, 19–32.

———. (1994). "Seeking Directions for Change: The Personalisation and Politicisation of Science Education." *Curriculum Studies* 2:1, 71–98.

Hoy, D. C., ed. (1986). *Foucault: A Critical Reader*. Oxford and New York: Basil Blackwell.

Hrdy, S. B. (1977). *The Langurs of Abu: Female and Male Strategies of Reproduction*. Cambridge, MA: Harvard University Press.

———. (1981). *The Woman That Never Evolved*. Cambridge, MA: Harvard University Press.

Huaco, G. A. (1999). *Marx and Sociobiology*. Lanham, MD: University Press of America.

Hubbard, R. (1982). "The Theory and Practice of Genetic Reductionism: From Mendel's Laws to Genetic Engineering" in Rose, S. P. R., ed. *Towards a Liberatory Biology*. London and New York: Allison and Busby.

Hubbard, R., and E. Wald (1993). *Exploding the Gene Myth: How Genetic Information Is Produced and Manipulated by Scientists, Physicians, Employers, Insurance Companies, Educators, and Law Enforcers*. Boston: Beacon Press.

Hull, D. L. (1988). *Science as a Process: An Evolutionary Account of the Social and Conceptual Development of Science*. Chicago: University of Chicago Press.

Hungerford, H. R., W. J. Bluhm, T. L. Volk, and J. M. Ramsey, eds. (1998). *Essential Readings in Environmental Education*. Champaign, IL: Stipes Publishing Company.

Janov, A. (2000). *The Biology of Love*. Amherst, NY: Prometheus Books.

Jones, G. (1980). *Social Darwinism and English Thought: The Interaction Between Biological and Social Theory*. Brighton, Sussex: Harvester Press.

Kac, E. (2005). *Telepresence and Bio Art: Networking Humans, Rabbits and Robots*. Ann Arbor, MI: University of Michigan Press.

Kalfoglou, A., J. Scott, and K. Hudson. (2005). "PGD Patients' and Providers' Attitudes to the Use and Regulation of Preimplantation Genetic Diagnosis." *Reproductive Biomedicine Online*. 11:4, 486–496.

Kaplan, G. (1994). "'Human Nature': Nazi Views on Jews and Women" in Tobach, E., and B. Rosoff, eds. *Challenging Racism and Sexism: Alternatives to Genetic Explanations*. New York: The Feminist Press at the City University of New York, 188–210.

Kaplan, G. and L. Rogers (2003). *Gene Worship: Moving Beyond the Nature/Nurture Debate over Genes, Brain and Gender*. New York: Other Press.

Kay, L. E. (1993). *The Molecular Vision of Life: Caltech, the Rockefeller Foundation, and the rise of the new biology*. Oxford: Oxford University Press.

Kaye, H. L. (1986). *The Social Meaning of Modern Biology: From Social Darwinism to Sociobiology*. New Haven: Yale University Press.

Keeton, W. T. (1967). *Biological Science*. New York: Norton.

Keller, A. G. (1915). *Societal Evolution: A Study of the Evolutionary Basis of the Science of Society*. New York: Macmillan.

Keller, E. F. (1983). *A Feeling for the Organism: The life and work of Barbara McClintock*. San Francisco: W.H. Freeman.

———. (1985). *Reflections of Gender and Science*. New Haven and London: Yale University Press.

———. (1992a). *Secrets of Life, Secrets of Death: Essays on Language, Gender, and Science*. New York: Routledge.

———. (1992b). "Nature, Nurture and the Human Genome Project" in Kevles, D. J., and L. E. Hood, eds. *The Code of Codes: Scientific and Social Issues in the Human Genome Project*. Cambridge, MA: Harvard University Press, 281–299.

———. (1994). "Master Molecules" in Cranor, C. F., ed. *Are Genes Us? The Social Consequences of the New Genetics*. New Brunswick, NJ: Rutgers University Press, 89–98.

———. (1995). *Refiguring Life: Metaphors of Twentieth-Century Biology*. New York: Columbia University Press.

———. (2000). *The Century of the Gene*. Cambridge, MA: Harvard University Press.

Keller, E. F., and E. A. Lloyd, eds. (1992). *Keywords in Evolutionary Biology*. Cambridge, MA: Harvard University Press.

Kellert, S., H. Longino, and C. K. Waters. (2006). *Scientific Pluralism*. Minneapolis: University of Minnesota Press.

Kevles, D. J., and L. E. Hood, eds. (1992). *The Code of Codes: Scientific and Social Issues in the Human Genome Project*. Cambridge, MA: Harvard University Press.

Kilani, Z., and L. Haj Hassan. (2001). "Sex Selection and Preimplantation Genetic Diagnosis at the Farah Hospital." *Reproductive Biomedicine Online* 4:1, 68–70.

Kipnis, N. (1998). "A History of Science Approach to the Nature of Science: Learning Science by Rediscovering It" in McComas, W. F., ed. *The Nature of Science in Science Education*. Norwell, MA: Kluwer Academic Publishers, 177–196.

Kitano, H., ed. (2001). *Foundations of Systems Biology*. Cambridge, MA: MIT Press.

Kitcher, P. (1987). *Vaulting Ambition: Sociobiology and the Quest for Human Nature*. Cambridge, MA: MIT Press.

———. (2001). *Science, Truth, and Democracy*. New York: Oxford University Press.

Klipp, E., R. Herwig, A. Kowald, C. Wierling, and H. Lehrach. (2005). *Systems Biology in Practice*. Weinheim, GDR. : Wiley-VCH.

Koslowski, P., ed. (1999). *Sociobiology and Bioeconomics: The Theory of Evolution in Biological and Economic Theory*. Berlin and New York: Springer.

Kragh, H. (1998). "Social Constructivism, the Gospel of Science and the Teaching of Physics." *Science and Education* 7:3, 265–294.

Kuhn, T. S. (1962). *The Structure of Scientific Revolutions*. Chicago: University of Chicago Press.

Laclau, E., and C. Mouffe. (1985). *Hegemony and Socialist Strategy: Towards a Radical Democratic Politics*. London: Verso.

Lakatos, I., ed. (1968). *The Problem of Inductive Logic*. Amsterdam: North-Holland.

Latour, B. (1987). *Science in Action: How To Follow Scientists and Engineers Through Society*. Cambridge, MA: Harvard University Press.

Latour, B. (1988). *The Pasteurization of France*. Cambridge, MA: Harvard University Press.

Latour, B., and S. Woolgar. (1979). *Laboratory Life: The Social Construction of Scientific Facts*. Beverly Hills: Sage Publications.

Lederman, N. (1986). "Students' and Teachers' Understanding of the Nature of Science: A Reassessment." *School Science and Mathematics* 86:2, 91–99.

———. (1992). "Students' and Teachers' Conceptions of the Nature of Science: A Review of the Research." *Journal of Research in Science Teaching* 29:4, 331–359.

———. (1995). "Suchting on the Nature of Scientific Thought: Are We Anchoring Curricula in Quicksand?" *Science and Education* 4, 371–377.

Lederman, N., P. Wade, and R. L. Bell. (1998). "Assessing Understanding of the Nature of Science: A Historical Perspective" in McComas, W. F., ed. *The Nature of Science in Science Education*. Norwell, MA: Kluwer Academic Publishers, 331–350.

Lerner, R. (2004). "Genes and the Promotion of Positive Human Development: Hereditarian Versus Developmental Systems Perspectives" in Coll, C. G., E. L. Bearer, and R. Lerner, eds. *Nature and Nurture: The Complex Interplay of Genetic and Environmental Influences on Human Behavior and Development*. Mahwah, New Jersey: Lawrence Erlbaum Associates Inc.

Levine, J. S., and K. R. Miller (1994). *Biology: Discovering Life* 2nd ed. Lexington, MA: D. C. Heath.

Levins, R., and R. C. Lewontin (1985). *The Dialectical Biologist*. Cambridge, MA: Harvard University Press.

Lewontin, R. (1991). *Biology as Ideology: The Doctrine of DNA*. Concord, ON: Anansi.

———. (1994). *Inside and Outside: Gene, Environment and Organism*. Worcester, MA: Clark University Press.

———. (2000a). *It Ain't Necessarily So: The Dream of the Human Genome and Other Illusions*. New York: New York Review Books.

———. (2000b). *The Triple Helix: Gene, Organism, and Environment*. Cambridge, MA: Harvard University Press.

Lewontin, R. C., S. P. R. Rose, and L. J. Kamin (1984). *Not in Our Genes: Biology, Ideology, and*

Human Nature. New York: Pantheon Books.

Lewontin, R., and R. Levins. (2007). *Biology Under the Influence: Dialectical Essays on Ecology, Agriculture and Health*. New York: Monthly Review Press.

Lindzey, G. (1969). "Genetics and the Social Sciences" in Manosevitz, M., G. Lindzey, and D. D. Thiessen, eds. *Behavioral Genetics: Method and Research*. New York: Appleton-Century-Crofts, 3–14.

Lippman, A. (1991). "Prenatal Genetic Testing and Screening: Constructing Needs and Reinforcing Inequities." *American Journal of Law and Medicine* 17, 15–50.

Livingstone, D. L. (1995). "For Whom 'The Bell Curve' Tolls." *The Alberta Journal of Educational Research* XLI:3, 335–341.

Longino, H. E. (1990). *Science as Social Knowledge: Values and Objectivity in Scientific Inquiry*. Princeton, NJ: Princeton University Press.

Longino, H. E. (2000). "Towards an Epistemology for Biological Pluralism" in Creath, R., and J. Maienschein, eds. *Biology and Epistemology*. Cambridge, UK, and New York: Cambridge University Press, 261–286.

Longino, Helen. (2002). *The Fate of Knowledge*. Princeton, NJ: Princeton University Press.

Lopreato, J. (1984). *Human Nature and Biocultural Evolution*. London and Boston: Allen & Unwin.

Lopreato, J., and T. A. Crippen (1999). *Crisis in Sociology: The Need for Darwin*. New Brunswick, NJ: Transaction Publishers.

Losos, J., K. Mason, and S. Singer. (2008). *Biology*. 8th ed. (Based on the work of P. Raven and G. M. Johnson.) New York: McGraw-Hill Higher Education.

Loving, C. C. (1987). "From the Summit of Truth to Its Slippery Slopes: Science Education's Journey Through Positivist-Postmodern Territory." *American Educational Review Journal* 34 (Fall), 42–452.

Low, B. S. (2000). *Why Sex Matters: A Darwinian Look at Human Behavior*. Princeton, NJ: Princeton University Press.

Lumsden, C. J., and E. O. Wilson (1981). *Genes, Mind, and Culture: The Coevolutionary Process*. Cambridge, MA: Harvard University Press.

———. (1983). *Promethean Fire: Reflections on the Origin of the* Mind. Cambridge, MA: Harvard University Press.

Lynch, M. (1985). *Art and Artifact in Laboratory Life*. London: Routledge and Kegan Paul.

Lynch, M. (1991). "Recovering and Expanding the Normative: Marx and the New Sociology of Scientific Knowledge." *Science, Technology and Human Values* 16: 2 (Spring), 233–248.

Lynch, M.(1992). "Going Full Circle in the Sociology of Knowledge: Comment on Lynch and Fuhrman." *Science, Technology and Human Values* 17:2 (Spring), 228–233.

Lynch, M. (1993). *Scientific Practice and Ordinary Action: Ethnomethodology and Social Studies of Science*. Cambridge, UK and New York: Cambridge University Press.

Lynch, M. P. (2001). *The Nature of Truth: Classic and Contemporary Perspectives*. Cambridge, MA: MIT Press.

Maasen, S. (1997). "Female Organisms' Orgasms: The Social and Biological Constitution of a 'Natural Phenomenon'" in Weingart, P., S. D. Mitchell, P. J. Richerson, and S. Maasen, eds. *Human by Nature: Between Biology and the Social Sciences*. London and Mahwah, NJ: Lawrence Erlbaum Associates, Publishers, 87–100.

MacDonald, D. (1996). "Making Both the Nature of Science and Science Subject Matter Explicit Intents of Science Teaching." *Journal of Science Teacher Education* 7:3, 183–196.

Macdonald, R. D. (2000). *An Exploration of Computer-Simulated Evolution and Small Group Discussion of Pre-Service Science Teachers' Perceptions of Evolutionary Concepts*. Unpublished PhD dissertation, Toronto.

Macpherson, C. B. (1964). *The Political Theory of Possessive Individualism: Hobbes to Locke*. Oxford: Oxford University Press.

Mannheim, K. (1936). *Ideology and Utopia*. New York: Harcourt, Brace and World.

Manosevitz, M., G. Lindzey, and D. D. Thiessen, eds. (1969). *Behavioral Genetics: Method and Research*. New York: Appleton-Century-Crofts.

Maranto, G. (1996). *Quest for Perfection: The Drive to Breed Better Human Beings*. New York: Scribner.

Marchak, M. P. (1988). *Ideological Perspectives on Canada*. 3rd ed. Toronto: McGraw-Hill Ryerson.

Marx, K., and F. Engels (1976). *The German Ideology*. 3d rev. ed. Moscow: Progress Publishers.

Marx, K. (1977). *Capital: A Critique of Political Economy* (Fowkes, B., Trans.). New York: Vintage Books.

Masters, R. D. (1982). "Is Sociobiology Reactionary? The Political Implications of Inclusive Fitness Theory." *Quarterly Review of Biology* 57, 275–292.

Matthews, M. R. (1992). "History, Philosophy, and Science Teaching: The Present Rapprochement." *Science and Education* 1, 11–47.

———. (1994). *Science Teaching: The Role of History and Philosophy of Science*. New York: Routledge.

———. (1997). "Introductory Comments on Philosophy and Constructivism in Science Education." *Science and Education* 6, 5–14.

———. (1998a). "Introduction and Foreword" in McComas, W. F., ed. *The Nature of Science in Science Education*. Norwell, MA: Kluwer Academic Publishers, xi–xxi.

———. (1998b). "In Defense of Modest Goals When Teaching About the Nature of Science." *Journal of Research in Science Teaching* 35:2, 161–174.

Maturana, H., and F. J. Varela, F. J. (1998). *The Tree of Knowledge: The Biological Roots of Human Understanding*. Revised ed. Boston and London: Shambhala Publications, Inc.

Maynard Smith, J. (1964). "Group Selection and Kin Selection." *Nature* 201, 1145–1147.

———. (1988). *Games, Sex and Evolution*. New York and London: Harvester.

———. (1989). *Did Darwin Get It Right? Essays on Games, Sex, and Evolution*. New York: Chapman and Hall.

Mayr, E. (1982). *The Growth of Biological Thought: Diversity, Evolution, and Inheritance*. Cambridge, MA: Belknap Press.

Mazur, A. (1981). "Media Coverage and Public Opinion on Scientific Controversy." *Journal of Communication* (Spring 1981), 106–115.

McComas, W. F. (1998a). "The Principal Elements of the Nature of Science: Dispelling the Myths" in McComas, W. F., ed. *The Nature of Science in Science Education*. Norwell, MA: Kluwer Academic Publishers, 53–70.

———. (1998b). "A Thematic Introduction to the Nature of Science: The Rationale and Content of a Course for Science Educators" in McComas, W. F., ed. *The Nature of Science*

in Science Education. Dordrecht, Boston, London: Kluwer Academic Publishers, 211–222.
———. ed. (1998c). *The Nature of Science in Science Education*. Dordrecht, Boston, London: Kluwer Academic Publishers.
McComas, W. F., M. P. Clough, and H. Almazroa (1998). "The Role and Character of the Nature of Science in Science Education" in McComas, W. F., ed. *The Nature of Science in Science Education*. Norwell, MA: Kluwer Academic Publishers, 3–39.
McComas, W. F., and J. K. Olson (1998). "The Nature of Science in International Science Education Standards Documents" in McComas, W. F., ed. *The Nature of Science in Science Education*. Norwell, MA: Kluwer Academic Publishers, 41–52.
McKinnon, S. (2005). *Neo-Liberal Genetics: The Myths and Moral Tales of Evolutionary Psychology*. Chicago: Prickly Pear Press.
McKinnon, S., and S. Silverman, eds. (2005). *Complexities: Beyond Nature and Nurture*. Chicago: The University of Chicago Press.
Merton, R. K. (1970). *Science, Technology and Society in Seventeenth-Century England*. New York: H. Fertig.
———. (1973). *The Sociology of Science: Theoretical and Empirical Investigations*. Chicago: University of Chicago Press.
Miller, K. R. (1999). *Finding Darwin's God: A Scientist's Search for Common Ground Between God and Evolution*. New York: Cliff Street Books.
———.(2008). *Only a Theory: Evolution and the Battle for America's Soul*. New York: Viking.
Miller, L. G. (1978). "Fated Genes" in Caplan, A. L., ed. *The Sociobiology Debate*. New York, Hagerstown, San Francisco, London: Harper and Row, 269–279. Originally published as "Fated Genes." *Journal of the History of the Behavioral Science* (April 1976), 183–190.
Montagu, A. (1956). *The Biosocial Nature of Man*. New York: Grove Press.
———. ed. (1980). *Sociobiology Examined*. Oxford and New York: Oxford University Press.
Moody, D. E. (1996). "Evolution and the Textbook Structure of Biology." *Science Education*, 80:4, 395–418.
Mulder, M. B., A. Maryanski, and J. Turner (1997). "Biology and Anthropology" in P. Weingart, P., S. Mitchell, P. Richerson, and S. Maasen, eds. *Human by nature: Between Biology and the Social Sciences*. Mahwah, NJ: Lawrence Erlbaum Associates, 31–38.
Mulkay, M. (1979). *Science and the Sociology of Knowledge*. London, Boston and Sydney: George Allen and Unwin.
Mumford, L. (1963). *Technics and Civilization*. New York: Harcourt Brace and World.
———. (1967). *The Myth of the Machine: Technics and Human Development*. New York: Harcourt Brace and World.
Nelkin, D., and M. S. Lindee (1995). *The DNA Mystique: The Gene as a Cultural Icon*. New York: Freeman.
Neuman, L. W. (2000). *Social Research Methods: Qualitative and Quantitative Approaches*. 4th ed. Boston, London, Toronto, Sydney, Tokyo and Singapore: Allyn and Bacon.
Nickels, M. K., C. E. Nelson, and J. Beard. (1996). "Better Biology Teaching by Emphasizing Evolution and the Nature of Science." *The American Biology Teacher*, 58:6, 332–336.
Noble, D. (2006). *The Music of Life: Biology Beyond The Genome*. Oxford: Oxford University Press.
Nola, R. (1997). "Constructivism in Science and in Science Education: A Philosophical

Critique." *Science and Education*, 6:1–2, 55–83.

Nott, M. (1994). "Teaching Physics and the Nature of Science Together: A Case Study." *Physics Education* 29, 170–176.

Oldroyd, D. (1986). *The Arch of Knowledge: An Introductory Study of the History of the Philosophy and Methodology of Science*. New York and London: Methuen.

Oyama, S. (1985). *The Ontogeny of Information: Developmental Systems and Evolution*. Cambridge, London and New York: Cambridge University Press.

———. (2000). *Evolution's Eye: A Systems View of the Biology-Culture Divide*. Durham, NC, and London: Duke University Press.

Parkes, A. S., and J. M. Thoday, eds. (1968). *Genetic and Environmental Influences on Behaviour: A Symposium Held by the Eugenics Society in September 1967*. New York: Plenum Press.

Paul, D. B. (1984). "Eugenics and the Left." *Journal of the History of Ideas* 45 (October 1984), 567–590.

———. (1991). "The Rockefeller Foundation and the Origins of Behavior Genetics" in Benson, K. R., J. Maienschein, and R. Rainger, eds. *The Expansion of American Biology*. New Brunswick, NJ: Rutgers University Press, 262–283.

———. (1994). "Eugenic Anxieties, Social Realities, and Political Choices." in Cranor, P. F., ed., *Are Genes Us? The Social Consequences of the New Genetics* (pp. 142–154). New Brunswick, NJ: Rutgers University Press.

———. (1995). "Textbook Treatments of the Genetics of Intelligence." *The Quarterly Review of Biology*, 60(3), 317–326.

———. (1998). *The Politics of Heredity: Essays on Eugenics, Biomedicine, and the Nature-nurture Debate*. Albany: State University of New York Press.

Pedretti, E., H. McLaughlin, R. Macdonald, and W. Kiyjinji (1998). *Exploring the Culture and Practice of Science Through Science Centres*. Paper presented at the Canadian Society for Studies in Education, Ottawa, Canada.

Pedretti, E., McLaughlin, H., Macdonald, R. & Gitari, W., (2001). "Visitor perspectives on the nature and practice of science: Challenging beliefs through A Question of Truth." *Canadian Journal of Science, Technology and Mathematics Education*, 1:4, 399-418.

Phillips, D. C. (1997). "Coming to Terms with Radical Social Constructivisms." *Science and Education* 6:1–2, 85–104.

Pigliucci, M. (2001). *Phenotypic Plasticity: Beyond Nature and Nurture*. Baltimore: Johns Hopkins University Press.

Pinker, Steven. (2002). *The Blank Slate: The Modern Denial of Human Nature*. New York: Viking.

Polanyi, M. (1966). *The Tacit Dimension*. Garden City, NY: Routledge and Kegan Paul.

Popper, K. (1963). *Conjectures and Refutations: The Growth of Scientific Knowledge*. London: Routledge and Kegan Paul.

Pozzer-Ardenghi, L., and W. Roth. (2004). "Students' Interpretation of Photographs in High School Biology Textbooks." Paper presented at the Annual Meeting of the National Association for Research in Science Teaching, Vancouver, B.C. (April 1–4).

Putnam, H. (1981). *Reason, Truth and History*. Cambridge: Cambridge University Press.

Raven, P. H., and G. B. Johnson (1995). *Understanding Biology*. 3rd ed. Dubuque, IA: William C. Brown.

———. (1999). *Biology*. 5th ed. New York: William C. Brown/McGraw-Hill.

Restivo, S. (1995). "The Theory Landscape in Science Studies: Sociological Traditions" in Jasanoff, S., G. Markle, J. Petersen, and T. Pinch, eds. *Handbook of Science and Technology Studies*. Thousand Oaks, CA, London, New Delhi: Sage Publications, 95–110.

Richards, R. J. (1987). *Darwin and the Emergence of Evolutionary Theories of Mind and Behavior*. Chicago and London: The University of Chicago Press.

Ridley, M. (2003). *Nature via Nurture: Genes, Experience and What Makes Us Human*. Toronto: HarperCollins.

Rifkin, J. (1998a). *The Biotech Century: Harnessing the Gene and Remaking the World*. New York: Jeremy P. Tarcher/Putnam.

———. (1998b). "The Sociology of the Gene." *Phi Delta Kappan* 79:9, 648–654.

Rose, H. (1994a). *Love, Power and Knowledge: Towards a Feminist Transformation of the Sciences*. London: Polity.

———. (1994b). "The Two-Way Street: Reforming Science Education and Transforming Masculine Science" in Solomon, J., and G. Aikenhead, eds. *STS Education: International Perspectives on Reform*. New York: Teachers College Press, 155–166.

Rose, H., and S. P. R. Rose (1982). "On Oppositions to Reductionism" in Rose, S. P. R., ed. *Against Biological Determinism*. London and New York: Allison and Busby, 50–59.

———. eds. (2000). *Alas, Poor Darwin: Arguments against Evolutionary Psychology*. London: Jonathan Cape.

Rose, S. P. R. (1980). "'It's Only Human Nature': The Sociobiologist's Fairyland" in Montagu, A., ed. *Sociobiology Examined*. New York and Oxford: Oxford University Press, 158–170. Originally published as "'It's Only Human Nature': The Sociobiologist's Fairyland." *Race and Class* 20:3 (1979).

———. ed. (1982a). *Against Biological Determinism*. London and New York: Allison and Busby.

———. ed. (1982b). *Toward a Liberatory Biology*. London and New York: Allison and Busby.

———. (1992). *The Making of Memory*. London: Bantam.

———. (1995). "The Rise of Neurogenetic Determinism." *Nature* 373, 380–382.

———. (1997). *Lifelines: Biology, Freedom, Determinism*. London: A. Lane.

Ross, A., ed. (1996). *Science Wars*. Durham, NC and London: Duke University Press.

Rothman, B. K. (1998). *Genetic Maps and Human Imaginations: The Limits of Science in Understanding Who We Are*. New York: W.W. Norton and Company.

Ruse, M. (1978). "Sociobiology: A Philosophical Analysis" in Caplan, A. L., ed. *The Sociobiology Debate*. New York, Hagerstown, San Francisco, and London: Harper and Row, 355–375.

———. (1979). *Sociobiology: Sense or Nonsense?* Dordrecht, Holland and Boston: D. Reidel.

———. (1996a). *But Is It Science? The Philosophical Question in the Creation/Evolution Controversy*. Amherst, NY: Prometheus Books.

———. (1996b). *Monad to Man: The Concept of Progress in Evolutionary Biology*. Cambridge, MA: Harvard University Press.

———. (1999). *Mystery of Mysteries: Is Evolution a Social Construction?* Cambridge, MA: Harvard University Press.

———. (2000). *The Evolution Wars: A Guide to the Debates*. Santa Barbara, CA: ABC-CLIO.

Sahlins, M. (1977). *The Use and Abuse of Biology: An Anthropological Critique of Sociobiology*. London: Tavistock Publications.

Sahotra, S. (2004) "From the *Reaktionsnorm* to the Evolution of Adaptive Plasticity: A Historical

Sketch from 1909–1999" in DeWitt, T. J., and S. M. Scheiner, eds. *Phenotypic Plasticity: Functional and Conceptual Approaches*. New York: Oxford University Press.

Saletan, W. (2007). "The Embryo Factory: The Business Logic of Made-to-Order Babies." *Washington Post*. Newsweek Interactive Co. LLC.

Sapp, J. (1987). *Beyond the Gene: Cytoplasmic Inheritance and the Struggle for Authority in Genetics*. New York: Oxford University Press.

———. (1990a). *Where the Truth Lies: Franz Moewus and the Origins of Molecular Biology*. Cambridge, UK, and New York: Cambridge University Press.

———. (1990b). "The Nine Lives of Gregor Mendel" in Le Grand, H. E., ed. *Experimental Inquiries*. Norwell, MA.: Kluwer Academic Publishers, 137–166.

Schifellite, C. (1980). *Hegemony, Consciousness and Education in Social Change*. Unpublished master's thesis, University of Toronto.

———. (1987). "Beyond Tarzan and Jane Genes: Towards a Critique of Biological Determinism" in M. Kaufman, ed., *Beyond Patriarchy: Essays by Men on Pleasure, Power, and Pain*. Toronto: Oxford University Press, 45–63.

———. (2002). "Professing Modest Claims in Education." *In Professing Education* 1:1, 9–11.

———. (2008). "Critical Criminology and 21st-Century 'Liberal' Eugenics." *The Critical Criminologist* 18:2, 13–17.

———. (2011). "10 Faces of Innovation for Engaging Future Engineers: Using Epistemological Challenges to Teach About Modest Epistemologies." *Journal of Engineering Education* 100:1, 48–88.

Segerstråle, U. C. O. (2000a). *Defenders of the Truth: The Battle for Science in the Sociobiology Debate and Beyond*. Oxford and New York: Oxford University Press.

———. (2000b). "Anti-Antiscience: A Phenomenon in Search of an Explanation: Part I: Anatomy of Recent 'Antiscience' Allegations" in Segerstråle, U., ed. *Beyond the Science Wars: The Missing Discourse About Science and Society*. Albany: State University of New York Press, 75–100.

Shapin, S. (1994). *A Social History of Truth: Civility and Science in Seventeenth-Century England*. Chicago: University of Chicago Press.

———. (1995). "Here and Everywhere: Sociology of Scientific Knowledge." *Annual Review of Sociology* 21, 289–333.

Singer, P. (2000). *A Darwinian Left: Politics, Evolution and Cooperation*. New Haven, CT: Yale University Press.

Slezak, P. (1998). "Sociology of Scientific Knowledge and Scientific Education: Part I" in Matthews, M. R., ed. *Constructivism in Science Education: A Philosophical Education*. Norwell, MA: Kluwer Academic Publishers, 159–188. Originally published as "Sociology of Scientific Knowledge and Scientific Education: Part I." *Science and Education* 3:3 (1994), 265–294.

Smith, D. E. (1987). *The Everyday World as Problematic: A Feminist Sociology*. Toronto: University of Toronto Press.

———. (1990). *The Conceptual Practices of Power: A Feminist Sociology of Knowledge*. Toronto: University of Toronto Press.

———. (1999). *Writing the Social: Critique, Theory, and Investigations*. Toronto: University of Toronto Press.

Smith, M. U., and L. C. Scharmann. (1999). "Defining Versus Describing the Nature of Science:

A Pragmatic Analysis for Classroom Teachers and Science Educators." *Science Education* 83, 493–509.

Sociobiology Study Group of Science for the People (1978). "Sociobiology: Another Biological Determinism" in Caplan, A. L. ed. *The Sociobiology Debate*. New York, Hagerstown, San Francisco and London: Harper and Row, 280–290. Originally published as "Sociobiology: Another Biological Determinism." *BioScience* 26:3 (1976).

Solberg, B. (2005). "The Concept of Selection: When Are You selecting? Is It Discriminatory?" in Jonsdottir, I., ed. *PGD and Embryo Selection: Report from an International Conference on Preimplantation Genetic Diagnosis and Embryo Selection*, 91–98.

Solomon, J. (1991). "Teaching About the Nature of Science in the British National Curriculum." *Science Education* 75:1, 95–103.

———. (1993). *Teaching Science, Technology and Society*. Philadelphia: Open University Press.

———. (1994a). "Conflict Between Mainstream Science and STS in Science Education" in Solomon, J., and G. Aikenhead, eds. *STS Education: International Perspectives on Reform*. New York: Teachers College Press, 3–10.

———. (1994b). "Knowledge, Values, and the Public Choice of Science Knowledge" in Solomon, J., and G. Aikenhead, eds. *STS Education: International Perspectives on Reform*. New York: Teachers College Press, 99–110.

Solomon, J., and G. Aikenhead, eds. (1994). *STS Education: International Perspectives on Reform*. New York: Teachers College Press.

Spencer, L. T. (1991). "Changes in the Way Sociobiology Has Been Covered in General Biology Textbooks." Presented at the Biannual Meeting of The International Society for the Study of the History, Philosophy and Sociology of Biology. Northwestern University, Evanston Illinois.

Spuhler, J. N., ed. (1967). *Genetic Diversity and Human Behavior*. Chicago: Aldine Publishing Company.

Steen, W. v. d., and W. Voorzanger. (1984). "Sociobiology in Perspective." *Journal of Evolution* 13, 25–32.

Stein, R. (2007). "The *Washington Post* 'Embryo Bank' Stirs Ethics Fears: Firm Lets Clients Pick Among Fertilized Eggs." *The Washington Post*, Saturday, January 6; Page A01. http://www.washingtonpost.com/wp-dyn/content/article/2007/01/05/AR2007010501953.html

Stern, C., and E. R. Sherwood, eds. (1966). *The Origin of Genetcs: A Mendel Sourcebook*. San Francisco: Freeman.

Suchting, W. A. (1992). "Constructivism Deconstructed." *Science and Education* 1:3, 223–254.

———. (1997). "Reflections on Peter Slezak and the 'Sociology of Scientific Knowledge.'" *Science and Education* 6:1–2, 151–195.

Sullenger, K., and S. Turner (1998). "Nature of Science: Implications for Education: An Undergraduate Course for Prospective Teachers" in McComas, W. F., ed. *The Nature of Science in Science Education*. Dordrecht, The Netherlands; Boston; and London: Kluwer Academic Publishers, 243–254.

Taylor, P. C., and W. W. Cobern (1998). "Towards a Critical Science Education" in Cobern, W. W., ed. *Socio-Cultural Perspectives on Science Education: An International Dialogue*. Nowell, MA: Kluwer Academic Publishers, 203–207.

Tobach, E., B. Rosoff, and M. Fooden, eds. (1994). *Challenging Racism and Sexism: Alternatives to Genetic Explanations*. New York: Feminist Press at the City University of New York.

Tobin, K., and C. J. McRobbie. (1997). "Beliefs About the Nature of Science and the Enacted Curriculum." *Science and Education* 6, 355–371.

Trenn, T. B. (1979). "Preface" in Fleck, L. *Genesis and Development of a Scientific Fact* (Bradley, F., and T. J. Trenn, Trans.). Chicago and London: University of Chicago Press, xiii–xix.

Trenn, T. B., and R. K. Merton. (1979). "Descriptive Analysis" in Fleck, L. *Genesis and Development of a Scientific Fact* (Bradley, F., and T. J. Trenn, Trans.). Chicago and London: University of Chicago Press, 154–165.

Trivers, R. L. (1971). "The Evolution of Reciprocal Altruism." *Quarterly Review of Biology* 46, 35–57.

———. (1985). *Social Evolution*. Menlo Park, CA: Benjamin Cummings.

Tuana, N., ed. (1989). *Feminism and Science*. Bloomington: Indiana University Press.

Turner, J. H., A. M. Maryanski, and B. Giesen, B. (1997). "Biology and Sociology" in Weingart, P., S. D. Mitchell, P. J. Richerson, and S. Maasen, eds. *Human by Nature: Between Biology and the Social Sciences*. London and Mahwah, NJ: Lawrence Erlbaum Associates, 19–31.

Tyson-Bernstein, H. (1988). *A Conspiracy of Good Intentions: America's Textbook Fiasco*. Washington, DC: The Council for Basic Education.

van den Berghe, P. L. (1979). *Human Family Systems: An Evolutionary View*. New York: Elsevier.

———. (1981). *The Ethnic Phenomenon*. New York: Elsevier.

van der Dennen, J., D. Smillie and D. R. Wilson, eds. (1999). *The Darwinian Heritage and Sociobiology*. Westport, CT: Praeger.

Van Rooy, W. (1999). "Controversial Biological Issues: An Exploratory Tool for Assessing Teacher Thinking in Relation to Classroom Practice." Paper presented at the 72nd Annual Meeting of the National Association for Research in Science Teaching, Boston, March 28–31.

Vandermeer, J. H. (1996). *Reconstructing Biology: Genetics and Ecology in the New World Order*. New York: Wiley.

Varela, F. J. (1979). *Principles of Biological Autonomy*. New York and Oxford: North Holland.

von Cranach, M., K. Foppa, W. Lepenies, and D. Ploog, eds. (1979). *Human Ethology: Claims and Limits of a New Discipline: Contributions to the Colloquium Sponsored by the Werner-Reimers-Stiftung für Anthropogenetische Forschung*. Cambridge, UK: Cambridge University Press.

von Glasersfeld, E. (1989). "Cognition, Construction of Knowledge and Teaching." *Synthese* 80(1), 121–140.

———. (1992)."Constructivism Reconstructed: A Reply to Suchting." *Science and Education* 1(4), 379–384.

Walker, K. A., D. L. Zeidler, M. L. Simmons, and W. A. Ackett. (2000). "Multiple Views of the Nature of Science and Socio-Scientific Issues." Paper presented at the Annual Meeting of the American Educational Research Association, New Orleans, LA, April 24–28.

Wallace, H. M. (2005). "The Misleading Marketing of Genetic Tests." *GeneWatch*, 18:2, 3-5.

Walsh, A. (1995). *Biosociology: An Emerging Paradigm*. Westport, CT: Praeger.

Washburn, S. L. (1978). "Human Behavior and the Behavior of Others." *American Psychologist* 33:5 (1978).

Webster, G., and B. Goodwin (1996). *Form and Transformation: Generative and Relational Principles in Biology*. Cambridge, New York and Melbourne: Cambridge University Press.

Weingart, P. (1997). "Preface" in Weingart, P., S. D. Mitchell, P. J. Richerson, and S. Maasen, eds. *Human by Nature: Between Biology and the Social Sciences*. Mahwah, NJ: Lawrence Erlbaum Associates, vii–x.

Weingart, P., S. Maasen, and U. C. O. Segerstråle (1997). "Shifting Boundaries Between the Biological and the Social: The Social and Political Contexts" in Weingart, P., S. D. Mitchell, P. J. Richerson, and S. Maasen, eds. *Human by Nature: Between Biology and the Social Sciences*. London: Lawrence Erlbaum Associates, 65–102.

Weingart, P., S. D. Mitchell, P. J. Richerson, and S. Maasen, eds. (1997). *Human by Nature: Between Biology and the Social Sciences*. Mahwah, NJ, and London: Lawrence Erlbaum Associates.

Weingart, P., U. C. O. Segerstråle (1997). "From 'Social Biology' to 'Sociobiology'" in Weingart, P., S. D. Mitchell, P. J. Richerson, S. Maasen, eds. *Human by Nature: Between Biology and the Social Sciences*. London: Lawrence Erlbaum Associates, 68–87.

Williams, G. C. (1966). *Adaptation and Natural Selection*. Princeton, NJ: Princeton University Press.

Williams, R. (1976). *Keywords: A Vocabulary of Culture and Society*. New York: Oxford University Press.

Wilson, E. O. (1975). *Sociobiology: The New Synthesis*. Cambridge, MA: Belknap Press of Harvard University Press.

———. (1978). "Foreword" in Caplan, A. L., ed. *The Sociobiology Debate*. New York, Hagerstown, San Francisco, and London: Harper and Row, xi–xiv.

———. (1980). *Sociobiology: The Abridged Edition*. Cambridge, MA, and London: Belknap Press of Harvard University Press.

———. (1982). *On Human Nature*. Toronto, New York, London and Sydney: Bantam Books.

———. (2000). *Sociobiology: The New Synthesis*. 25th anniversary ed. Cambridge, MA: Belknap Press of Harvard University Press.

Wright, R. (1994). *The Moral Animal: Evolutionary Psychology and Everyday Life*. 1st ed. New York: Pantheon Books.

Young, R. M. (1985). *Darwin's Metaphor*. Cambridge, New York, New Rochelle, Melbourne, and Sidney: Cambridge University Press.

Zachary, G. P. (2000). "Global Sperm Trade a Fertile Business: Denmark Finds Niche in World Market." *Wall Street Journal*. Friday, January 7.

Zirkle, C. (1951). "Gregor Mendel and His Precursors." *Isis* 42, 97–104.

Zuckerman, H. (1988). "The Sociology of Science" in Smelser, N. J., ed. *Handbook of Sociology*. Newbury Park, CA: Sage Publications, Inc., 511–574.

INDEX